Politics of Urban Runoff

Urban and Industrial Environments

Series editor: Robert Gottlieb, Henry R. Luce Professor of Urban and Environmental Policy, Occidental College

Politics of Urban Runoff

Nature, Technology, and the Sustainable City

Andrew Karvonen

The MIT Press
Cambridge, Massachusetts
London, England

© 2011 Massachusetts Institute of Technology

For information about special quantity discounts, please email special_sales@mitpress .mit.edu.

This book was set in Sabon by Graphic Composition, Inc., Bogart, Georgia. Printed and bound in the United States of America.

Library of Congress Cataloging-in-Publication Data

Karvonen, Andrew, 1971–
Politics of urban runoff : nature, technology, and the sustainable city / Andrew Karvonen.
   p.   cm.—(Urban and industrial environments)
Includes bibliographical references and index.
ISBN 978-0-262-01633-9 (hardcover : alk. paper)—ISBN 978-0-262-51634-1 (pbk. : alk. paper)
1. Urban runoff. 2. Water quality management—Political aspects. 3. Water quality—Texas—Austin. 4. Water quality—Washington (State)—Seattle. I. Title.
TD657.K374   2011
363.72'84—dc22

2011003569

10 9 8 7 6 5 4 3 2 1

Contents

Preface

Water flows through the urban fabric and leaves the trace of its power behind it.
—André Guillerme[1]

It's raining in the city. Dark clouds release raindrops that fall steadily through the atmosphere and pick up airborne pollutants as they descend to the landscape below. The drops make a cacophony of small thuds as they land on the various surfaces of the city—rooftops, roads, trees, sidewalks, and lawns. The water collects in puddles at first and then transforms into temporary streams of stormwater runoff, picking up terrestrial pollutants from automobiles, lawn fertilizers and pesticides, pet droppings, errant trash, and so on. The stormwater becomes a torrent as it flows into gutters and underground pipe networks before being discharged into a local waterbody. The storm eventually abates, the puddles and waterways dry up and disappear, and the city prepares for the rain cycle to begin anew.

The mundane act of rainfall in the city serves as the starting point for this study. Rain produces a particular mood in cities—shiny surfaces and spotted windows, puddles with oil rainbows, flowing water in gutters, rushing creeks—but is also the cause of serious and confounding problems including flooding, erosion, landslides, and water pollution. As U.S. cities developed in the nineteenth and twentieth centuries, municipal governments and private property owners created impervious landscapes of transportation networks and building rooftops, reorienting stormwater flows from predominantly vertical to horizontal pathways. Managing these new surface flows became the responsibility of technical professionals, namely municipal engineers, who developed a logic of efficient conveyance to remove stormwater volumes from cities as quickly as possible to protect human lives and property as well as the flows of commerce.

The logic of efficient conveyance came under fire in the 1960s and 1970s with the rise of contemporary environmentalism and the recognition that the changed hydrologic cycle in urban areas impacts the ecological health of receiving waterbodies. These impacts are particularly evident in urban creeks and streams that receive large, erosive stormwater volumes that alter predevelopment conditions dramatically. In the most extreme cases, waterways have been completely channelized or buried to separate nature from the built environment. Today, despite four decades of regulatory and enforcement activity in the United States, urban runoff persists as a confounding problem for technical experts and municipal officials who attempt to balance flood control and erosion with water quality, aesthetics, and community character.

Stormwater runoff is most often discussed in terms of large technical networks, bureaucratic management, and engineering and scientific expertise, but it is also implicated in a wide range of cultural, economic, and political issues of urban life. In the last four decades, technomanagerial actors have been joined by ecologists and biologists, landscape architects and ecological designers, and environmental activists and local residents to address the impacts of urban runoff. Today, issues of urban runoff are implicated in debates and activities aimed at producing improved urban futures that are more sustainable, resilient, and livable. More broadly, urban runoff is part of the larger debate over the dilemma of urban nature and the tensions between nature, technology, and humans in cities.

To examine the implications of urban runoff, I adopt a relational perspective that is informed by environmental history, landscape theory, human and cultural geography, ecological planning, and science and technology studies.[2] Scholars in these disciplines use the topic of urban nature to reveal the messy, hybrid relations among nature, society, and technology while contemplating how these relations can be reworked to address the tensions between humans and nature. Here, it is understood that "cities are dense networks of interwoven sociospatial processes that are simultaneously human, material, natural, discursive, cultural, and organic."[3] In this study, urban runoff serves as a lens to unpack the dilemma of urban nature, to explore the various ways that it is shaped through urban development activities, and to suggest avenues of collective action to realize improved urban futures.

The aims of the study are threefold. The first is to follow in the footsteps of William Cronon, Matthew Gandy, Erik Swyngedouw, Maria Kaika, and others by presenting a detailed account of the cultural and political implications of urban nature. Second, the study strengthens the connections between urban environmental history, landscape and ecological design, and the sociotechnical study of cities by highlighting the importance of technical networks, expert knowledge, and bureaucratic regulation in the governance of urban nature. And third, the study connects notions of hybrid natures from urban ecology with notions of politics, governance, and citizenship as forwarded by scholars of environmental politics and ecological democracy.[4] Thus, it addresses not only how we think about urban nature but also how we collectively act upon it.

I begin in chapter 1 by summarizing the historic evolution of urban runoff practices, focusing on the dominant role of engineers and bureaucratic actors in the management of urban hydrologic flows. The evolution of practice from urban drainage to stormwater management goes hand in hand with the formalization of urban governance in the early twentieth century as well as the emergence of environmental management in the 1960s and 1970s. In chapter 2, I contrast the dominant engineering approach with that of landscape architects and ecological planners, and argue that a relational perspective opens up urban nature to more nuanced interpretations. Rather than forwarding a compartmentalized view of urban nature, a relational perspective recognizes the indelible connectedness of urban residents with their material surroundings. Such a perspective is helpful to understand how the contemporary conditions of cities came into being and how they can be reworked into more desirable configurations.

The following four chapters are case studies of urban runoff activities in Austin, Texas, and Seattle, Washington—two U.S. cities in which urban water flows have been a central topic of political debate over the last four decades. The cities have heightened public dialogs on urban runoff and are also recognized by practitioners for their exemplary municipal stormwater management programs.[5] The prominence of urban runoff both publically and professionally in each city provides a rich trove of ideas, events, and actors to contemplate how water flows are implicated in the functioning of the contemporary city. Furthermore, the urban runoff activities in these cities are part of wider discussions and debates of sustainable urban development. Both Austin and Seattle are considered to be leaders in sustainability, and urban water activities feed into this reputation.[6]

The case studies comprise empirical data collected through qualitative methods—namely semistructured interviews and textual analyses—to highlight prominent stormwater activities in each city with a particular emphasis on historical, cultural, and political implications.[7] To complete the case studies, I conducted thirty-four semistructured interviews between 2006 and 2008 with actors involved in urban runoff activities, including local and regional government officials, design professionals and natural scientists, and community and environmental activists (see table 0.1). I granted these individuals anonymity to elicit unguarded opinions about the activities in question and to protect them from possible negative repercussions for participating in the study.

I also completed textual analyses in each city composed of primary and secondary materials including engineering and planning reports, maps, photographs, plans, local history accounts, academic dissertations and theses, journals, newspapers, magazines, and websites. In Austin, I relied on the library collections at the University of Texas at Austin and the City of Austin, particularly the Austin History Center. In Seattle, I studied documents in the University of Washington's general and special collections as well as the Seattle Public Library and the Museum of History and Industry. Based on the plethora of textual sources (particularly municipal reports and newspaper articles), I quickly realized that obtaining a complete picture of stormwater activities in each city would be an impossible task. Rather, the goal of reviewing the textual and visual documents was to identify key actors and activities to understand how urban runoff is associated with processes of urban development and the evolving cultural and political character of each city. The resulting study is not intended to be a comprehensive environmental history of Austin and Seattle but rather a series of examples that illustrate how urban runoff activities—and, by extension, urban nature—are important to the past, present, and future of each city.

The Austin case studies, presented in chapters 3 and 4, center on a culturally relevant community space, Barton Springs. Since the 1970s, contentious public debates about protecting the water quality of the springs have resulted in highly restrictive land use regulations upstream as well as an ambitious municipal program to purchase land for nature conservation. Urban runoff in Austin highlights the tensions between urban development, property rights, comprehensive land use planning, and citizen activism. The upstream activities also have implications downstream of

Table 0.1
Summary of interview subjects by group

	Austin	Seattle
Local and regional government staff members	6	8
Design professionals and natural scientists	3	6
Environmental and community activists	7	4

Barton Springs, as land development restrictions to protect environmental quality have pushed development activities to the urban core. Environmental justice activists in these downstream areas have exposed the conflicts between environmental protection and social equity, particularly where existing low-income and minority communities have been targeted for redevelopment. As a whole, the Austin case studies exemplify the challenge of mitigating environmental impacts while encouraging economic growth and forwarding goals of social equity in a fragile and rapidly urbanizing landscape.

In chapters 5 and 6, I present the Seattle case studies. Seattle has a long-standing reputation as a city in harmony with nature due to its location in the lush, green Pacific Northwest. However, historical evidence reveals a multitude of interventions by the municipal government to reorient the existing landscape to facilitate economic and population growth. The seemingly "natural" landscape is in fact a product of significant human alteration. The romantic view of Seattle's landscape persisted until the 1950s when public recognition of declining water quality in Lake Washington led Seattle residents and the government to embrace new approaches to human/nature relations. Today, a multitude of activities by the municipality and community activists aims to restore and protect urban waterways to improve water quality, reintroduce salmon populations, reduce environmental management costs, and foster community cohesion. The Seattle case studies suggest that the popular notion of a city in harmony with nature has inspired a whole host of interventions with unanticipated implications on human/nature relations.[8]

In chapter 7, I compare and contrast the various historical and contemporary activities of urban runoff in Austin and Seattle by introducing a framework of environmental politics. "Politics" is defined here in broad terms to encompass activities by both governmental and nongovernmental actors to alter the relationship between urban residents and their material surroundings. I argue that the municipal government typically engages in rational politics that make use of technical experts and bureaucratic actors to manage urban nature. Meanwhile, neighborhood and civil society groups participate in populist political activities to challenge the rational political decisions of the municipal government and open up urban nature to alternative voices. Rational and populist politics exist in a feedback loop, with residents serving as a check on municipal government activities. As an alternative to the rational/populist feedback loop, I present examples of civic politics that engage residents and local government in participatory, action-oriented, and constructive practices to reinterpret and reorient urban hydrologic flows. Civic politics recognize that water flows and society are co-constitutive, mirroring the relational perspective of urban nature in chapter 2, and use this interconnectedness to deliberate and act upon urban nature to produce more desirable conditions.

In the final chapter, I make three suggestions to further the relational perspective of urban nature by fostering civic political practices. The transition to a relational politics of urban nature can be facilitated by developing *civic imaginaries* to inspire new connections between humans and nonhumans, by developing *civic expertise* to mediate and translate this relational perspective, and by conducting *civic experiments* to test and refine this relational approach. The intent here is not to produce a political atmosphere with a deep-seated environmental ethic but rather to foster an attitude toward urban nature that is situated, inclusive, and action-oriented.

Urban nature continues to be a topic of significant debate and disagreement, despite major advances in technological development, ecological science, and democratic governance over the last century. Urban runoff serves as a proxy to study the tensions between humans and nature as well as amongst humans by opening up technomanagerial practices of environmental management to scrutiny. Geographer Erik Swyngedouw summarizes these tensions: "The political and the technical, the social and the natural, become mobilized through socio-spatial arrangements that shape distinct geographies and landscapes; landscapes that celebrate the visions of the elite networks, reveal the scars suffered by the disempowered and nurture the possibilities and dreams for alternative visions."[9] Ultimately, it is these dreams of alternative visions that fuel new conceptions of urban nature as well as activities to rework human/nature relations. By recognizing our embeddedness in the urban landscape and by taking action to modify our relations with both humans and nonhumans, we can realize improved urban futures, and in the process, better understand our place in the world.

Acknowledgments

I am indebted to many individuals and organizations for assistance with the research and writing of this book. First and foremost, I thank my mentor and friend Steven Moore for his diligent guidance, constructive criticism, and unremitting support in my investigation of urban water politics. He reviewed numerous early drafts and provided helpful comments that inevitably improved my arguments. I am also grateful to Simon Guy and my colleagues at the Manchester Architecture Research Centre for providing a productive academic environment to finalize the manuscript.

I was fortunate to receive financial support from several organizations. The U.S. National Science Foundation's Science and Society Program provided funding for my Seattle fieldwork under Grant No. 0646720 and the Texas Water Resources Institute at Texas A&M University provided funding for my Austin fieldwork through a partnership with the U.S. Geological Survey. The School of Architecture at the University of Texas at Austin served as my home base during my research period and served as a thought-provoking work environment. All opinions, findings, and conclusions or recommendations are mine alone and do not necessarily reflect the views of the U.S. National Science Foundation, the Texas Water Resources Institute, the U.S. Geological Survey, or the University of Texas at Austin.

My colleagues and friends were never short of suggestions and comments that collectively enhanced the final insights and arguments. I am particularly grateful to Paul Adams, Michael Barrett, Kent Butler, Bruce Hunt, and Fritz Steiner, scholars who generously devoted their time and ideas to this transdisciplinary study. I also benefited from conversations and debates with Julian Agyeman, Ralf Brand, Govert Geldof, Govind Gopakumar, David Hess, Andy Jamison, Jin-Oh Kim, Marty Melosi, Tim Moss, Liz Mueller, Dean Nieusma, Michael Oden, Jonathan Ogren, Lynn Osgood, Cameron Owens, Barbara Parmenter, Bob Paterson, Gert de Roo,

Bjørn Sletto, Peter Stahre, Erik Swyngedouw, Barbara Wilson, Langdon Winner, Ned Woodhouse, Ken Yocom, and Ming Zhang.

With respect to empirical data gathering, I relied on anonymous respondents in Austin and Seattle to provide insights on urban runoff practices and debates. Their unvarnished perspectives were crucial to unpacking the multiple urban water issues in these cities and I am grateful to them for sharing their ideas and opinions with me. A number of individuals in Austin and Seattle helped me identify these respondents as well as relevant textual data sources. I extend my appreciation to Denise Andrews, Kevin Autry, Jon Beall, Jackie Chuter, Cory Crocker, Matt Hollon, Jack Howison, Shawn Kavon, David Levinger, Ryan Robinson, and Lauren Ross. I would be remiss if I did not acknowledge the contributions of my nonhuman respondents: the various waterways, pipes, fish populations, treatment systems, and vegetation that communicated with me in various ways during my field investigations. None of these actors consented to participation in the study, but I sincerely hope that I represented them accurately.

And finally, I thank my wife, Sonya, for her enduring patience and sympathetic ear.

1

The Dilemma of Water in the City

[The contemporary city has] a tendency to loosen the bonds that connect its inhabitants with nature and to transform, eliminate, or replace its earth-bound aspects, covering the natural site with an artificial environment that enhances the dominance of man and encourages an illusion of complete independence from nature.
—Lewis Mumford[1]

The illusion of independence from nature is the dominant perspective of cities today not only for the general public but also for the majority of urban theorists and practitioners. The city is understood as the antithesis of nature, a refuge from the untamed and uncivilized hinterland, a place wholly constructed by and for human habitation. The idealized modern city is human-created, rational, permanent, and ultimately devoid of unruly nature.[2] This idea of the domination of nature by humans was also central to the nineteenth-century notion of Progress, an idea that can be reduced to a simple formula: "Progress equals the conquest of nature by culture."[3] The allure of Progress and urbanization conspired to create a cognitive split between premodern and modern, countryside and city, nature and culture.

Technological development also played a central role in separating the city from the countryside. As historian David Nye argues, "Nineteenth-century Americans repeatedly told themselves stories about the mastery and control of nature through technology in which radical transformations of the landscape were normal developments."[4] The shift from the nineteenth-century organic city to the twentieth-century rational city has much to do with the introduction of large technical systems for water, sewer, transportation, communication, and electricity services.[5] The large-scale, centralized, and permanent character of these systems was significant in shaping contemporary forms of municipal governance and also in establishing the practice of comprehensive city planning.[6] These systems not only required long-range vision on the part of municipal officials for population growth forecasting, technical network design, and infrastructure financing, but also called for a permanent bureaucracy to administer them continuously.[7] Political scientist John Dryzek describes this as a process of "administrative rationalism" wherein environmental problems were resolved through the harnessing of scientific expertise by the administrative state.[8] As a whole, these systems are sociotechnical amalgams of hardware (pipes, roads, pumps, transfer stations, etc.) and software (municipal bureaucracies, technical experts, urban residents, codes and regulations, etc.) that constitute an idealized urban metabolism.[9]

In this chapter, I examine a particular form of urban water, stormwater runoff, to understand how technical experts were tasked with taming urban nature through technological intervention. I begin with a brief overview of the evolution of urban runoff and emphasize the role of late nineteenth-century engineers and public health officials who advocated for an increase in the hydrologic metabolism of cities to improve sanitary conditions. A well-drained city allowed for economic development to flourish while also creating sanitary conditions that improved the health of urban residents. The logic of urban runoff shifted in the 1960s and 1970s from a drainage approach that emphasized the efficient conveyance of water away from the city as quickly as possible to a stormwater management approach that involved slowing down and treating water volumes to address issues of water quality, erosion, and flooding.[10] The evolution of urban runoff networks also contributed to the rise of local and regional governments while situating technomanagerial actors as the central arbiters of human/nature relations in cities.

The Promethean Project and the Control of Nature

The control and rationalization of the urban landscape in the nineteenth and twentieth centuries involved a multitude of activities to tame and control nature through technology, human labor, and capital investment. Geographer Maria Kaika identifies three phases of this so-called Promethean Project of Modernity.[11] The first phase, the Nascent Promethean Project, began in the early nineteenth century as advocates of urban development attributed deteriorating social and environmental conditions to an unruly and undisciplined natural landscape. During this period, technical experts and public health officials called for ambitious, far-reaching plans to control and subdue nature for human ends and subsequently solidified their central role in urbanization processes. The second phase, the Heroic Moment of Modernity's Promethean Project, lasted from roughly the late nineteenth to the first three quarters of the twentieth century and was a period of large-scale construction of infrastructure networks. It was during this period that large technical systems would come to dictate the form and function of contemporary cities, serving as the material and administrative backbone of urban development. And the third phase, Modernity's Promethean Project Discredited, began in the last decades of the twentieth century and continues to the present day. This period is characterized by increasing demand for resources as well as crumbling infrastructure and environmental disasters. Here, it is understood that nature is no longer tamed but rather a source of crisis, with humans exerting widespread detrimental impacts on nature.[12]

The Promethean Project is a simplified narrative of the changing relations between nature, technology, and humans. Taken too literally, it could be interpreted as a teleological notion of societal progress with humans slowly becoming "enlightened" about the negative consequences of employing technology to control nature. But the Promethean Project serves as a helpful heuristic framework to describe the dominant narrative of human progress and the resulting dilemma of nature. It highlights the central role of technology and technical experts in the mediation of human/ nature relations in the context of nineteenth- and twentieth-century urban development and is particularly apt for understanding the process of urbanization as a means to rework nature for human ends.[13]

The Engineer as Champion of the Promethean Project

Undoubtedly, the hero of the Promethean Project is the engineer who became "the modern Prometheus, the One Man who would stand alone against nature."[14] The profession of engineering emerged in the first half of the nineteenth century as self-trained experts were increasingly tapped to resolve the nature/culture dilemma.[15] Well into the twentieth century, the engineer was widely understood as the harnesser of nature for human ends, as illustrated by the 1913 proclamation of a chemical engineer:

What is Engineering? The control of nature by man. Its motto is the primal one— "Replenish the earth and subdue it." . . . Is there a barren desert—irrigate it; is there a mountain barrier—pierce it; is there a rushing torrent—harness it. Bridge the rivers; sail the seas; apply the force by which all things fall, so that it shall lift things. . . . Nay, be "more than conqueror" as he is more who does not merely slay or capture, but makes loyal allies of those whom he has overcome! Appropriate, annex, absorb, the powers of physical nature into human nature![16]

This quotation from a 1913 issue of the *Journal of Industrial and Engineering Chemistry* epitomizes the arrogance and boldness of engineering ambition at the turn of the twentieth century.[17] The work of engineers was central to Progressive Era reform efforts in which it was understood that improved material conditions of cities would lead to improved moral and physical health of urban residents. As Kaika argues, "The heroes of modernity promised to dominate nature and deliver human emancipation employing imagination, creativity, ingenuity, romantic heroic attitude, and a touch of hubris against the given order of the world."[18] Engineers simultaneously positioned themselves as a technical elite that would save society through the application of scientific and empirical knowledge while also purporting to be apolitical and efficient problem solvers.[19] Urban historians Stanley Schultz and Clay McShane note that "engineers professed to work above the din of local politics" and became the dominant actors in municipal government:

Engineers stamped their long-range visions of metropolitan planning on the public consciousness over the last half of the nineteenth century. Their successful demands for political autonomy in solving the physical problems of cities contributed to the ultimate insistence for efficient government run by skilled professionals. At the heart of physical and political changes in the administration of American cities, indeed at the very core of city planning, stood the work of the municipal engineers.[20]

Creating Modern Urban Water Metabolisms

In addition to the engineer, water was and continues to be a central actor in the pursuit of societal progress, the rollout of large technical systems, and the form and function of the contemporary city. As geographer Matthew Gandy writes, "Water is not simply a material element in the production of cities but is also a critical dimension in the social production of space."[21] Three types of urban water networks would be central to modernization efforts in the nineteenth century: water supply, sanitary sewerage, and drainage. The first two networks are tightly linked through the introduction of consistent and plentiful water supplies and the subsequent development of the water-carriage form of sanitary waste disposal. These technical networks would provide sanitary services for urban residents and emerge as the hydrologic circulatory system of the contemporary city.[22]

The third form of water network, drainage, has received less attention from urban environmental historians because the systems do not embody the technological wonder of water supply and sanitary sewers and also because these networks were and continue to be an amalgam of human and natural, built and unbuilt. Rather than a wholly new network of technical components, drainage networks consist of a jumble of natural and technical elements (ditches, existing waterways, natural depressions, streets, rooftops, downspouts, and so on) that defy tidy description. Furthermore, drainage networks do not fit nicely in the Promethean timeline because they are a premodern form of infrastructure that preceded the nineteenth-century water supply and wastewater networks. As early as 3000 BC, human settlements had extensive drainage networks to convey rainwater away from buildings and streets. The celebrated Roman Empire, best known for its aqueducts and roads, had equally impressive drainage networks that were central to its success.[23] These practices were largely abandoned after the fall of the Empire and urban drainage strategies tended to be informal and piecemeal with ditches or gutters made of wood or stone to convey urban runoff to the nearest waterway until the mid-nineteenth century. With the advent of the engineering profession as well as the technical and administrative means to design and construct complex service networks, underground storm drains gradually emerged as the standard method to convey urban runoff in North America and Northern Europe.[24]

The contemporary distinction between the three urban water networks reflected in engineering education as well as construction and administration activities belies a messier history of development during the nineteenth century. Illicit dumping of sanitary wastes in surface drainage networks caused widespread public health and aesthetic problems. Furthermore, the introduction of consistent water supplies and the adoption of water-hungry plumbing fixtures (particularly the water closet) caused a breakdown in existing sanitary systems. Pits for the collection of human waste were overwhelmed with the rapid increase in wastewater volumes, resulting in widespread public health issues. In short, there was no systems view of urban water in the nineteenth century; water supply and sewerage were treated as separate operations throughout the century, while the increasing density of urban populations and the increasing consumption of water and generation of wastewater volumes created complex linkages between urban residents and water flows.[25]

In the 1850s, U.S. cities began to adopt water-carriage systems for sanitary wastes, transforming waste disposal from a local, labor-intensive effort into a capital-intensive but convenient method for urban residents to maintain more healthy urban conditions.[26] The adoption of water-carriage systems was justified by the prevailing theory of miasmas held by public health officials that disease was transmitted through gases emanating from the decay of waste. Although the theory was scientifically flawed, it emphasized sanitation as a primary focus of nineteenth-century city builders and called for the removal of sanitary wastes to avoid epidemics.[27] A clean city was a healthy city, and cleansing would be accomplished through the adoption of the water-carriage system to remove disease-causing wastes.

Urban drainage flows were implicated with the new water-carriage system for sanitary wastes through the construction of combined sewer systems, in which both sanitary and stormwater volumes were conveyed in a single pipe. For nineteenth-century engineering and sanitary experts, the combined sewer design was a cost-effective and efficient solution that avoided the buildup of miasmas that were thought to be produced by standing water.[28] However, other technical experts advocated for separated sewers in which sanitary wastes were carried in a smaller pipe and stormwater volumes were carried either in a larger pipe or in ditches on the surface. Engineers who promoted the separated sewer design argued that sanitary wastes contained valuable organics that could be recovered and used as agricultural fertilizer.

The choice of combined versus separate sewers was complicated by a number of factors including hydraulics, cost, design features, perspectives on disease transmission, and the influence of internationally recognized sewerage experts such as Colonel George E. Waring Jr. However, by the 1890s, most engineers concluded that both systems provided equivalent sanitary services, with combined systems being more appropriate in large cities where space for the surface conveyance of urban runoff was infeasible.[29] In essence, population density and economic development took precedence over hydrologic flows in large cities, which required water flows to be relegated underground. The variety of sewer networks first built in the nineteenth century would result in a hodgepodge of drainage networks, often within the same city, including surface networks, subsurface separated sewers, and subsurface combined sewers.[30] In many cases, these networks continue to be used today due to the high cost and disruption associated with upgrading them.

With respect to the Promethean Project, urban drainage would involve increasingly sophisticated hydrologic approaches to convey these water volumes to downstream receiving waterbodies. Water was interpreted as an unruly element in the urban landscape and the enclosed pipe was the primary strategy to discipline it.[31] As a whole, stormwater engineering

Figure 1.1
The Los Angeles River as the quintessential example of the Promethean Project applied to urban drainage. (Note that the project is not complete; the rationalized channel still includes vegetation and wildlife.)

has been dominated by rational, scientific approaches to size drainage networks and convey stormwater from cities as quickly as possible, a logic that can be summarized as "end of pipe, out of sight, out of mind."[32] The logic of technical efficiency is exemplified by the Los Angeles River, a 51-mile-long concrete channel built by the U.S. Army Corps of Engineers in the 1940s to tame an unruly waterbody (see figure 1.1).[33] Reflecting on the channel, Gandy notes, "The peculiar landscape of the river channel— its apparent emptiness and artificiality—reinforces a general perception that everything in Los Angeles has been artificially constructed."[34] As such, the rational approach to urban runoff with its emphasis on efficient conveyance entails particular assumptions about the relationship between nature and society.

Recognition of End-of-Pipe Pollution Problems

The increased efficiency of sewer networks to capture and expel wastes resulted in markedly improved sanitary conditions in cities but created unanticipated end-of-pipe pollution problems.[35] With all wastewater networks, the ultimate outlet was, and continues to be, local waterways. Engineers and public health officials of the nineteenth century believed that natural waters were self-purifying and that the discharge of sewage into waterways was thus a form of wastewater treatment. When this disease etiology was debunked in the late nineteenth century, engineers embraced the dilution principle—"dilution is the solution to pollution"—which used a mass balance argument to justify the deposit of untreated sewage into receiving waters.[36] In effect, natural waterbodies became implicated in the hydrologic metabolism of the contemporary city not only as a means of conveyance but as a form of passive wastewater treatment.

By 1905, most U.S. states had laws regarding water pollution, but these were intended to reduce local nuisances rather than prevent contamination of drinking water for downstream communities. Furthermore, only a handful of these states had adequate enforcement provisions in place.[37] There was frequent disagreement among sanitary experts over the need for wastewater treatment. Engineers advocated for water-treatment techniques for potable water supplies while arguing that dilution provided adequate treatment for wastewater volumes. Public health officials were quick to embrace germ theory when it was first proposed in the late nineteenth century and called for wastewater treatment to protect downstream cities and towns from waterborne diseases.[38] The growing recognition of bacteria as the primary disease-causing mechanism—along with litigation by downstream cities and towns over deteriorating water quality conditions—would eventually lead to the adoption of treatment strategies for sanitary and combined sewer discharges. By World War II, the majority of populations in large U.S. cities were served by sewage treatment, and today, both water treatment and wastewater treatment are standard practices in almost all communities.[39]

The Environmental Era as a Challenge to the Promethean Project

With respect to urban drainage, the third phase of the Promethean Project began in the 1960s as biologists and ecologists recognized that urban runoff had significant amounts of pollutants that were detrimental to environmental and human health. Stormwater had traditionally been referred to as "whitewater" because it was assumed to be as clean as rain, but there was growing recognition among scientists, engineers, and public health officials that urban runoff was polluted and would require treatment.[40] As early as the 1930s, biologists and ecologists in Northern Europe and North America identified significant concentrations of contaminants in urban runoff, recognizing that rain falling through the atmosphere scrubbed airborne pollutants and picked up contaminants from surfaces such as sediment, hydrocarbons, microbial organisms, heavy metals, and toxic substances.[41] The efficient conveyance logic of urban drainage introduced in the nineteenth century and perfected throughout the twentieth century would increasingly come into conflict with calls for improved environmental quality in the 1960s. This would mark the shift from "urban drainage" to "stormwater management"—from getting rid of urban runoff as quickly as possible to containing and treating it before release to downstream waterbodies. In short, both the quantity and quality of stormwater would require management.

In the United States, widespread attempts to address the water quality impacts of urban runoff began with the passage of the Federal Water Pollution Control Act of 1972, later renamed the Clean Water Act (CWA). The objective of the CWA is succinct, explicit, and all-encompassing: "to restore and maintain the chemical, physical, and biological integrity of the Nation's waters."[42] The legislation is well known for its requirements to reduce and treat "point source pollution" from discrete sources such as industrial outfalls and wastewater treatment plants, but the CWA also addressed "nonpoint source" pollution—those diffuse volumes of wastewater from agricultural activities and urban runoff. Point source pollution is relatively easy to manage with conventional wastewater treatment strategies; nonpoint source pollution tends to elude treatment because of its dispersed, ubiquitous character as well as significantly larger volumes of wastewater requiring treatment (sometimes two orders of magnitude greater than point source volumes) and appreciably lower concentrations of contaminants.[43]

The enormous challenges of addressing nonpoint source pollution resulted in little progress in reducing or cleaning up pollution from urban runoff in the 1970s and 1980s. The most notable activity in stormwater research was an ambitious study of 2,300 precipitation events in twenty-eight major metropolitan areas by the U.S. Environmental Protection Agency (EPA) between 1977 and 1983 to characterize the environmental impacts of urban runoff.[44] Results from the National Urban Runoff Program were transformed into regulation via the 1987 amendments to the Clean Water Act with the formation of the National Pollutant Discharge Elimination System (NPDES) permitting program that established minimum standards for stormwater treatment.[45] The NPDES program first targeted larger metropolitan areas with populations of greater than one hundred thousand beginning in 1990 because of their administrative capacity to establish municipal stormwater programs, followed by smaller municipalities beginning in 1999.[46]

Despite the increasing recognition that the engineered drainage networks of the nineteenth century were the primary cause of nonpoint source pollution, engineers continued to be central actors in stormwater management. By the late 1960s, sanitary engineers had rebranded themselves as environmental engineers to reflect new societal concerns about environmental pollution and to expand their constituency. Historian Martin Melosi writes, "The almost blind faith in science and technology exhibited in years past was replaced by a more complex, and sometimes schizophrenic, relationship. The nation's triumphs in science and technology were sometimes blamed for the excesses of the new consumer culture. At the same time, the advice of scientists and engineers was sought out to help eradicate pollution and restore a more amenable quality of life."[47] The drainage networks designed and constructed by engineers were understood to be the cause of nonpoint source pollution and their successors were tasked with finding stormwater management solutions. Despite the growing ambivalence of the public to the increased application of science and technology, engineers have largely continued to engage in a Promethean approach to solving water quality problems by expanding and upgrading existing drainage infrastructure to include treatment facilities.[48] As Dryzek notes, "Prometheans have unlimited confidence in the ability of humans and their technologies to overcome any problems presented to them—including what can now be styled environmental problems."[49]

The Systemic Nature of Nonpoint Source Pollution

The extent of urban water degradation is not limited to pollutant concentrations, the major emphasis of federal regulatory requirements, but implicates the hydrologic cycle as a whole.[50] From a systems perspective, processes of urbanization redirect rainfall from infiltrating into the subsurface, creating surface runoff. Activities to redirect rainfall became particularly pronounced beginning in the early twentieth century, as asphalt and concrete pavements rapidly became the norm for urban transportation networks.[51] Durable pavements (as opposed to gravel and macadam materials) facilitated faster transportation to serve streetcar suburbs and commercial goods delivery, and created consistent surfaces and grades to protect subsurface sewer networks. Paving was an important element of sanitary reform in U.S. cities that also contributed to the conveyance logic of urban drainage.[52]

In the 1950s, local governments began to adopt urban street standards to accommodate rapid transmission of automobile traffic and to ensure

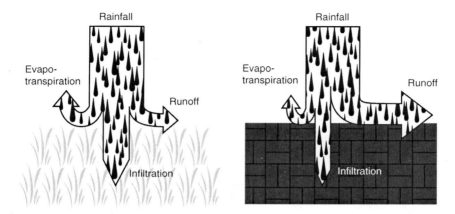

Figure 1.2
An illustration of preurban (left) and urban (right) rainfall flowpaths with urbanization resulting in reduced evapotranspiration and infiltration flows and increased surface runoff. Illustration by Shawn Kavon.

an aesthetically pleasing look for streets through specifications for curb-and-gutter design.[53] With these standards, the conveyance logic of the nineteenth century became embedded in the codes and regulations of city building and had both functional and cultural implications. Today, urban development typically involves the removal of vegetation and topsoil to allow for the construction of buildings, roads, and other infrastructure. Whereas an undeveloped parcel of land may infiltrate 90 percent of rainfall, urbanized areas have significantly less infiltrative capacity, resulting in larger volumes of runoff that change the hydrologic cycle from a largely vertical flowpath to a largely horizontal one (see figure 1.2).[54] Drainage networks are then installed to convey surface runoff to the nearest waterway for discharge. Overall, the built environment of U.S. cities consists of impervious surfaces that are about two-thirds transportation infrastructure and one-third building roofs.[55]

The rationalization of the urban landscape for water conveyance also had a number of implications for the biological and ecological function of watersheds, none of which were understood or even considered during the storied network-building period from the mid-nineteenth century to the 1960s. The increase in impervious surfaces results in a "flashy" hydrologic regime, with higher peak flows in receiving waterbodies that increase flooding and bank erosion processes, threaten habitat for vertebrate and invertebrate populations, and lower baseflows during dry periods due to depleted groundwater volumes (see figure 1.3). Stormwater picks up

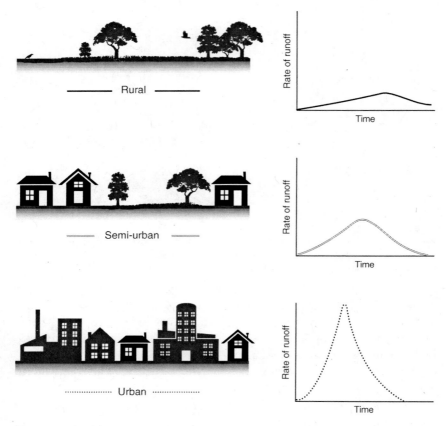

Figure 1.3
A comparison of stormwater runoff rates versus time for rural, semi-urban, and urban conditions. Urban areas experience intense runoff rates with high flows over short periods of time. Illustration by Shawn Kavon.

a variety of pollutants—fertilizers, pesticides, bacteria, sediment, nutrients, oils and greases, metals, organic matter, toxic chemicals, trash and debris—from rooftops, streets, and sidewalks and discharges them into waterways. Landscape theorist Michael Hough writes, "The benefits of well-drained streets and civic spaces are paid for by the costs of eroded stream banks, flooding, impaired water quality and the disappearance of aquatic life."[56] The logic of efficient conveyance of urban runoff is at odds with contemporary calls for environmental protection.[57]

Today, the most widely used strategy to manage stormwater volumes is detention ponds. Detention ponds were originally developed to control flooding in urban areas by detaining water temporarily and releasing it slowly, but these facilities also capture contaminants as stormwater volumes slow down and solids drop out. Detention ponds are a common element of cities, whether they are fenced-off concrete boxes in sprawling parking lots, artificial ponds behind buildings on private property, or grassy depressions in public rights-of-way. They have become the de facto strategy to manage the flooding, erosion, and water quality issues of urban runoff in many cities because they are relatively easy to construct and because they provide flood protection and a modest degree of water-quality treatment.

However, detention ponds are only partially effective as a water treatment strategy because they provide physical removal of contaminants but do not address those contaminants that remain suspended in solution. These suspended pollutants flow through detention ponds and are released downstream, where they have biological and ecological impacts. Furthermore, there are social costs associated with detention ponds, namely large land requirements (estimated at 0.5 to 2 percent of the catchment area), unattractive aesthetics, and public safety concerns.[58] Finally, there are significant challenges in maintaining these systems over time, from removing captured pollutants to controlling insects and rodents and monitoring to determine when they need to be replaced or upgraded.[59] In effect, detention ponds require a high level of management costs by local governments while providing only partial environmental protection.

Other stormwater management strategies involve a variety of so-called best management practices (BMPs) such as ponds, vaults, tanks, separators, swales, filter strips, wetlands, basins, perforated pipes, trenches, French drains, sand and biological filters, and porous pavements. These strategies are used to store, treat, and infiltrate urban runoff volumes to mitigate issues of flooding, soil erosion, and water pollution. The end-of-pipe treatment strategies of conventional stormwater management have a wide range of efficacy for pollutant removal and are not particularly reliable, despite the claims of product manufacturers and designers (see table 1.1).[60] Urban runoff continues to confound engineering solutions, although there has been an enormous amount of scientific research, pilot studies, regulations, construction, administration, and maintenance of these systems. In other words, nonpoint source pollution problems have largely eluded the Promethean promise of the complete control of nature.[61] At the end of the twentieth century, over half of the receiving water bodies in the United States did not meet their designated water uses and water quality goals as specified in the CWA.[62] It is tempting to

Table 1.1
Average pollutant removal rates as a percentage for common stormwater BMPs

Strategy	Total suspended solids	Total phosphorous
Dry extended detention pond	25 to 81	10 to 40
Wet basin	33 to 97	29 to 75
Stormwater wetland	43 to 93	16 to 68
Bioretention filter	75 to 95	61 to 83
Sand filter	68 to 96	15 to 77
Infiltration trench	90 to 100	30 to 100

Source: Weiss, Gulliver, and Erickson 2007, 226.
Note: Represents the 67 percent confidence interval. Some values are assumed.

blame technical experts and ineffective local governments for this failure, but the confounding character of nonpoint source pollution is a classic "wicked problem" of environmental protection—one that is not easily solved through technomanagerial activities.

One of the most significant barriers to effective stormwater management is that cities were already in place before water quality issues were raised. Thus, retrofitting has enormous costs because the embedded logic of efficient conveyance is difficult to reverse both materially and administratively.[63] Moreover, urban runoff is not amenable to conventional "command and control" forms of environmental management due to its dispersed and low concentration character.[64] It is a systemic problem that transcends technical and administrative rationality because it implicates a whole host of urban development activities, from impervious surface propagation to land use and zoning regulations, growth management, resident behavior, the location and character of vegetation, maintenance schedules, and so on. There is a fundamental mismatch between the end-of-pipe strategies that dominate conventional stormwater management and the complex character of urban runoff.

Conclusions

Historic and contemporary practices of urban drainage have contributed to the supposed separation of the city from nature. Engineers played a significant role in rationalizing urban water flows to allow for urban growth by constructing urban drainage networks that could convey stormwater as quickly as possible away from cities. The 1960s marked a turning point in urban runoff practices from a logic of efficient conveyance to focus on water quality and to address multiple goals of quantity and quality simultaneously. Despite four decades of scientific research and practice, urban runoff has largely eluded control by technological means. As a stormwater engineer notes, "The fundamental problem with conventional stormwater management may be the mindset. It does not treat water as a valuable resource but more like a problem to solve, or even worse, seeks to export it as a waste product."[65]

To address the systemic problem of urban runoff, there is a need to reflect on this idea that the city is somehow devoid of nature. Once we begin to understand the connections between water flows and human habitation, and between nature and cities, we can start to recognize urban runoff as an issue that transcends technomanagerial strategies and has profound implications on broader issues of urban development. In the next chapter, I examine the approach of landscape architects and ecological planners, a group of urban practitioners whose work on urban water has gone largely unacknowledged until recently. They have long forwarded an alternative to the Promethean approach of total control of nature and instead embraced different ideas of human/nature relations in the city. Such ideas recognize the systemic character of urban runoff and the importance of water flows in the production of urban space.

2
Urban Runoff and the City of Relations

It is not sufficient to understand either the processes of the social system or the processes of the natural system alone. Both mold the city's physical environment, which forms the common ground between them.

—Anne Whiston Spirn[1]

The Promethean narrative involving the technomanagerial control of urban nature is dominant in North America and Northern Europe today. However, other urban runoff logics have been forwarded to reform the Promethean approach or to embrace an entirely different interpretation of urban nature. Such practices recognize the folly in attempts to expunge nature from cities or to bring it under complete human control. In this chapter, I begin by describing the emergence of Low Impact Development and Sustainable Urban Drainage as a significant evolution in stormwater management. These approaches have the potential to be more effective at reducing the environmental impacts of urban runoff but continue to perpetuate technology and technomanagerial actors as the ultimate arbiters of human/nature relations. However, the rise of source control also opens up urban runoff to other practitioners, namely landscape architects and ecological designers, with a long but unheralded history in shaping urban water flows. These practitioners often embrace a relational perspective that rejects modernist dichotomies of city/rural, human/nonhuman, and culture/nature and instead focuses on the indelible connections between urban residents and their material surroundings. Such an approach has radical implications for stormwater practice and more broadly for conceiving and enacting different forms of urban nature.

Source Control as an Alternative Urban Drainage Logic

Beginning in the 1980s, a small but influential group of stormwater practitioners began to recognize that managing urban runoff would not be achieved through the command and control practices of traditional engineering.[2] Rather, there was a need to completely rethink the logic of urban drainage. Using a systems perspective from ecological science, these practitioners posited that the most effective way to manage urban runoff was by reducing the volumes generated at the source. Referred to as Low Impact Development (LID) in North America and Sustainable Urban Drainage Systems (SUDS) in Northern Europe, source control turns the urban drainage regime on its head by emphasizing the reduction of stormwater volumes and cleansing at the point of generation through site design, infiltration, and treatment strategies.[3] Where conventional stormwater management focuses on the symptoms of large stormwater volumes, source control goes to the root of the problem to address development patterns and impervious cover that create these large volumes of polluted water in the first place.[4] LID pioneer Larry Coffman writes that "the basic goal of LID is to engineer a site with as many small-scale retention, detention, prevention, and treatment techniques as needed to achieve the hydrologic functional equivalent to predevelopment conditions."[5]

Proponents of source control advocate for the judicious use of ecological science principles rather than physical strategies of conventional engineering. The approach has much in common with the relatively new field of ecological engineering, which adopts principles of ecological science to engineering work and shifts the emphasis of stormwater management from hydraulics to hydrology.[6] Rather than focusing on storage and treatment of stormwater volumes, source control uses distributed and decentralized management strategies such as bioswales, pervious paving, green roofs, and rainwater harvesting to mimic the predevelopment hydrologic cycle as closely as possible. This approach is promoted as a way not only to reduce environmental impacts but also to reduce construction and maintenance costs; it can be understood as a "green governance" approach to urban runoff issues for which the mantra is "pollution prevention pays."[7] The rise of source control as an alternative drainage strategy suggests that urban water flows should not be directed by environmental regulations focused on water quality and quantity but rather by a systems focus on the relationship between ecological flows and human habitation.

Today, source control represents a small but growing contingent of stormwater practitioners who are gradually reforming urban runoff practices to improve environmental quality. Through the implementation of pilot projects, the retrofitting of existing networks, and the revision of municipal codes, source control advocates are updating the Promethean approach by introducing a variety of new strategies that mimic nature while redefining the relations between water and urban residents. There is a growing consensus that source control will eventually become the dominant form of stormwater management, although this transition will likely occur over decades if not centuries due to the ubiquity of the conventional stormwater management logic in the built environment.

Source control represents a significant break from conventional stormwater management but technical experts continue to dominate, with engineers joined by biologists and ecologists. Further, municipal and regional governments continue to be the central arbiters of human/nature interactions; the traditional structure of environmental management remains firmly in place. As such, source control is more an evolution in stormwater management than a revolution, as many of its advocates would have us believe. The idea that natural systems and human systems are somehow separate continues to pervade the logic of source control, as does the Promethean faith in human ingenuity to overcome the problem of nature.

Landscape Architecture and Urban Water Flows

The emergence of source control as an alternative logic of stormwater management highlights the influence of another group of urban practitioners who are frequently overlooked in the history of urban water: landscape architects and ecological planners. Frederick Law Olmsted Sr. is often cited as the father of urban planning and landscape architecture; his work in the second half of the nineteenth century exemplifies an alternative conception of the relationship between urban residents and their surroundings.[8] Olmsted developed his design approach from earlier landscape gardening practitioners in Europe and the United States in the nineteenth century. He was also a prominent figure in the sanitary reform movement in the United States in the mid-nineteenth century; this movement emphasized a connection between the moral reform of urban residents and material conditions.[9] By the late nineteenth century, Olmsted had developed an organic philosophy for landscape architecture and urban planning that integrated the social and natural metabolisms of the urban environment.

Olmsted strove to differentiate the work of landscape architects from other experts of the built environment, namely engineers, architects, and horticulturists. He felt that the landscape architect had a "constructive imagination" that was frequently missing from those other disciplines, and he was particularly critical of engineering in this regard. He wrote, "A good Engineer is nearly always impatient of [the] indefiniteness, unlimitedness and mystery which is the soul of landscape."[10] In contrast,

Olmsted promoted a vision that was a synthesis of social, technical, and environmental design. From the English landscape gardening movement, he adopted an aesthetic of irregular and naturalistic landscape composition as opposed to the formal geometries of the Renaissance and baroque traditions. This natural aesthetic was a strategy to improve the physical and psychological well-being of city populations who were coping with increasing congestion and the loss of contact with the nonhuman world.

With respect to urban water, the Back Bay Fens project in Boston is a quintessential example of Olmsted's attempt to integrate urban residents with natural processes. Constructed in the 1880s and 1890s on tidal flats and floodplains polluted with industrial and residential sewage, the project included a buried sewer pipe, a parkway, and a streetcar line that allowed for the intermingling of human and nonhuman activities.[11] The design was a daring multifunctional experiment that involved landscape design, engineering, urban planning, and ecology to serve as infrastructure and recreational grounds. The project functioned as designed for only a few decades; in 1910, engineers completed the Charles River dam, negating the flood control and water quality aspects of Olmsted's project.[12]

From the late nineteenth century to the 1930s, city planners followed Olmsted's lead and applied an integrated landscape approach to urban development, notably with the Garden City, City Beautiful, and Regional Planning movements. These approaches emphasized the importance of public parks, botanical gardens, and tree-lined boulevards to maintain a connection between urban residents and nature. However, the integrated vision of city and nature would gradually fade in the twentieth century as disciplinary specialization resulted in a splintered approach to landscape design and the abandonment of Olmsted's synthetic vision of nature, technology, and society. As a result, landscape architects became associated with environmental artistry and urban planners defined themselves as rational and pragmatic professionals. Landscape theorist Anne Spirn argues that the marginalization of landscape architects as artists was a consequence of Olmsted's ability to successfully hide the unnatural parts of his designs. She writes: "He disguised the artifice so that ultimately, the built landscapes were not recognized and valued as human constructs. . . . Ironically, it was the 'natural' appearance of his work that prevented people from appreciating how it fulfilled a broad range of functions."[13]

Landscape theorist Catherine Howett echoes this critique, arguing that Olmsted's vision was "too complex and fragile for most Americans to understand."[14] Consequently, the integrated approach to urban water management forwarded by Olmsted and his successors would be an anomaly of twentieth century urban development rather than standard practice. Specialization of urban design and development would contribute to the modernist dichotomy of form versus function with aesthetic practices such as park design and landscaping relegated to landscape architecture, while the regulation of urban waterflows was the province of engineers. The emphasis of landscape architecture was first and foremost on aesthetics, leaving function and governance to engineers and urban planners.[15] This division of labor between form and function would have important implications for the governance of urban nature; Schultz notes that "unlike engineers, landscape architects contributed little to the reformulation of public policy or to changes in administrative reach and authority of municipal governments."[16]

The Emergence of Ecological Planning

At the beginning of the contemporary environmental era, landscape architects broke the mold of aesthetic practice and reembraced the integrated design approach developed by Olmsted.[17] After World War II, landscape architect Garrett Eckbo promoted landscape design and planning as a science for social engineering, and in the 1960s, ecological planner Ian McHarg crystallized the idea of merging landscape design with ecological science in his groundbreaking book *Design with Nature*.[18] However, McHarg rejected the organic philosophy of nineteenth-century landscape design in favor of a rational, objective practice based on the principles of ecological science. His approach involved the layering of different attributes atop one another to plan cities within existing environmental conditions, a strategy that would later be adapted to the widely used practices of geographic information science and environmental impact statements. McHarg's so-called layer-cake method continues to be favored by contemporary conservation biologists and landscape ecologists who use complex remote sensing and modeling techniques to describe the ecology of cities.[19]

Like Olmsted, McHarg was interested in the relationships between nature, society, and technology, but he replaced Olmsted's embrace of mystery and soul with a rational and objective approach based in the natural sciences. McHarg was highly critical of the artistic focus of landscape architecture that had dominated since the 1930s, noting that "art has occult and esoteric pretensions and an intrinsic obscurantism."[20] He emphasized landscape architecture and ecological planning practices as the application of natural science principles and advocated for the rejection of a signature artistic style in favor of collaborative interdisciplinary work.[21]

For McHarg, form followed function, and function was determined by the principles of ecological science. He recognized that the application of scientific principles was useful to garner influence, respect, and power, as did engineers in the nineteenth century.

One of McHarg's most famous projects has urban drainage as its structural backbone. The Woodlands, a twenty-seven-thousand-acre master planned community located thirty miles north of Houston, Texas, was built in the 1970s by oil and gas magnate George Mitchell. The project was part of a larger movement in the 1960s and 1970s to reimagine suburban development and metropolitan form; today, the Woodlands is often heralded as the most comprehensive example of McHarg's ecological planning method.[22] Along with his design team from the noted landscape architecture firm of Wallace, McHarg, Roberts & Todd, McHarg identified flooding as the most critical environmental concern of the project. The team used the natural contours of the site to create a surface drainage system that preserved floodplain areas, limited impervious surface coverage, and promoted infiltration of stormwater rather than conveyance and disposal, predating the source control strategies of LID practitioners by several decades (see figure 2.1).[23]

Today, critics of the Woodlands recognize its innovative aspects but frequently denounce it as a green and attractive form of suburban sprawl. Planning theorist Ann Forsyth characterizes the Woodlands as an "eco-burb" because the car-based development is shrouded in a lush forest of trees and natural waterways. The project comprises 16 percent open space and 4.4 dwellings per residential acre (equivalent to typical suburban residential density), and the most prominent visual feature of the Woodlands is its wooded aesthetic.[24] The implicit message of the Woodlands is that suburban development densities are required to protect the ecological integrity of the landscape and ecological science can be used to justify the appropriate density for human habitation. Reflecting on the Woodlands, urban theorists Andres Duany and David Brain write, "The outcome of this environmental model is that the better parts of nature are preserved while traffic-generating and socially dysfunctional development is camouflaged with a natural aesthetic."[25]

Duany and Brain's assessment reflects a larger critique of McHarg's ecological science approach to landscape architecture and site planning by those who emphasize the social rather than natural aspects of cities. McHarg's critics describe the layer-cake method as deterministic, utilitarian, and ecocentric; indeed, McHarg describes himself as an "ecological determinist" and his colleagues characterize him as an evangelist of the

Figure 2.1
The Woodlands development north of Houston, Texas, features a plethora of natural features in a suburban community

environment.[26] Landscape theorist James Corner argues that the scientific approach to urban ecology as forwarded by McHarg treats nature as an object and does not consider the attendant cultural and social issues; Howett adds that McHarg's approach is linear, logical, and comprehensible, leaving little room for mystery, eccentricity, and chaos.[27] At stake are the expressive and emotive attributes of landscape that are purportedly lost in the rational view of the ecological scientist.

Today, there is a palpable split in urban ecological practice and theory between those who conceive and shape complex ecological systems and those who attempt to create iconic landscapes as artistic expressions. Landscape theorist Louise Mozingo argues that in "the most skeptical extremes of this continuum, aesthetic exploration is trivial; ecological regimen is determinism not design."[28] Spirn, a student of McHarg and a central member of the team that developed the Woodlands project, laments the divergence of art and science between ecological planners and landscape architects, and is particularly critical of the specialized thinking employed by both camps. She writes:

Now pieces of landscape are shaped by those whose narrowness of knowledge, experience, values, and concerns leads them to read and tell only fragments of the story. To an ecologist, landscape is habitat, but not construction or metaphor. To a lawyer, landscape may be property to regulate, to a developer, a commodity to exploit, to an architect, a site to build on, to a planner, a zone for recreation or residence or commerce or transportation, or "nature preservation.". . . So each discipline and each interest group reads and tells landscape through its own tunnel vision of perception, value, tool and action.[29]

The disciplinary split is yet another result of the modernist distinctions between nature and culture, science and art, wilderness and city. The Promethean Project described in the previous chapter centered on the engineer as the primary actor in defining the distinction between city and nature, but this perspective of humans residing outside of, above, or separate from nature pervades all members of society. Landscape theorist Peggy Bartlett notes that urban culture "celebrates a sophisticated distance from the messy realities of farm and wilderness. As Woody Allen famously asserts, 'I am two with Nature.'"[30] The public understanding of cities is predicated upon the opposition between the constructed human world and the unconstructed natural world. This is the prevalent perspective today, despite an increasing understanding that cities are highly dependent on "nature's services" of air, water, nutrient flows, and sunlight.[31] Even environmental actors embrace this dualism of nature/culture when they advocate for preservation strategies that attempt to reduce the impact of society on nature by zoning society out of nature via urban growth boundaries and the preservation of natural areas.[32] The built environment of cities effectively serves as the social pole and wilderness serves as the natural pole.[33] Landscape theorist Elizabeth Meyer provides a helpful critique of this dualist form of thinking: "The continuation of the culture-nature and man-nature hierarchies by designers when they describe the theoretical and formal attributes of their work perpetuates a separation of humans from other forms of life, vegetal and animal. This separation places people outside the ecosystems of which they are a part and reinforces a land ethic of either control or ownership instead of partnership and interrelationship."[34]

This critique of a segmented understanding of landscape provides a helpful segue to introduce an alternative interpretation of urban nature. Unlike engineers of the nineteenth and twentieth century who extended the Promethean Project to address various urban runoff issues, an influential group of theorists and practitioners have continued to embrace the ambivalence of urban nature. They understand that cities do not represent a material liberation from biological necessity but rather that nature is a persistent and ever-changing presence in the built environment. In other words, "cities are built in nature, with nature, through nature."[35] In this regard, Spirn writes, "The realization that nature is ubiquitous, a whole that embraces the city, has powerful implications for how the city is built and maintained and for the health, safety, and welfare of every resident."[36]

Contemporary urban ecologists share a common perspective of urban nature as a particular arrangement of humans, nature, and technology rather than a strict replacement of the natural with the unnatural. From a geographic perspective, urbanization is not the "end of nature" as many would argue but rather a transformation into a different spatialization.[37] The principal argument of urban nature centers on the ontological division between humans and nature.[38] Geographers Bruce Braun and Joel Wainwright argue that what counts as nature should never be a closed question because speaking about nature with some level of certainty entails foreclosing on other possibilities.[39] Braun describes this ontological perspective as one that is defined not by immutable essences but rather as "precarious achievements, although no less consequential for being so."[40] Feminist scholar Donna Haraway describes the activity of questioning such seemingly self-evident ontological assumptions as a "queering" strategy that opens them up to critical scrutiny.[41]

Not surprisingly, the queering of seemingly self-evident categories of humans and nature induces the wrath of those who do not subscribe to an environmental ethic as well as those who see themselves as custodians of the wild.[42] The idea that nature no longer transcends the built environment but is entirely and inextricably implicated within it is antithetical to contemporary environmental activists who often attempt to reify the perceived boundaries between the human and nonhuman. Environmental historian William Cronon provides a helpful summary of contemporary urban ecology as the problem of nature or wildness: "If wildness can stop being (just) out there and start being (also) in here, if it can start being as humane as it is natural, then perhaps we can get on with the unending task of struggling to live rightly in the world—not just in the garden, not just in the wilderness, but in the home that encompasses them both."[43]

Cronon shares with urban ecologists an understanding that nature is not "out there" but is inextricably bound up with humans. It is this characteristic of binding or relating that is crucial to the activities of designers who intervene in the built environment. The goal of urban practitioners—including landscape architects, ecological planners, and engineers—should be to elucidate the relations between the natural, social, and technical and to base their designs on desirable relational configurations. The emphasis on relations is an ecological perspective, but it is one that is markedly different from the ecology of the natural sciences. Humans are situated within a complex array of heterogeneous relations in which nature and society are outcomes rather than causes; they are wrapped up in the co-construction of material and nonmaterial imbroglios. Here, nature does not merely serve as a backdrop for human activities or determine human existence but is instead an active participant in the making of human societies.[44] From this perspective, sociologist Bruno Latour argues that the importance of ecology is not its scientific certainty or its empirical data gathering practices but rather in its ability to relate elements of the world, stating that, "It is on these associations and not nature that ecology must focus."[45]

Relations are central to most if not all ecological discourses as well as theories of globalization, complexity, and chaos, among many others. Relational theory has become a particularly vibrant discourse in human and cultural geography and has resonance in landscape theory.[46] Over a century ago, John Muir wrote, "When we try to pick out anything by itself, we find it hitched to everything else in the universe."[47] It is this hitching process that is a central element of ecological or relational thinking.[48] Latour describes this process of relating as a new form of political activity, one that is not based on general theories grounded in social or natural imperatives but on the relationships between entities (the politics of relational thinking will be discussed in detail in chapter 7).[49] The relational interpretation of urban nature has the potential to make ecology a subversive subject, echoing Paul Sears's famous 1964 proclamation that if ecology were "taken seriously as an instrument for the long-run welfare of mankind, [then it would] endanger the assumptions and practices accepted by modern societies, whatever their doctrinal commitments."[50] Indeed, one can only imagine how the urban landscape would be different had the "modern Prometheus" instead been the "modern relationalist."

The relational perspective described previously is similar to Olmsted's landscape approach and is being revived as a guiding principle for contemporary landscape architecture and ecological planning in mediating natural and social flows. A relational perspective emphasizes processes over objects and subjects while rejecting the science/art dualism that pervades conventional urban ecological practice and theory. Landscape theorist Forster Ndubisi summarizes this position: "The underlying wisdom here is that we need to understand the character of the landscape not only in terms of its natural processes, but also in terms of the reciprocal relationship between people and the landscape. The important word is *relationships*."[51]

Hybrids, Relational Space, and Topology

Urban ecologists who forward a relational perspective recognize that cities are composed not of subjects and objects but of impure entities—hybrids or cyborgs—that are at once human and nonhuman, cultural and natural. There is a perpetual insistence on the "mongrel nature of the world."[52] Hybrids are not a social construction of relational theorists but are ever present in the world; we have simply chosen to ignore them and instead perceive the world in terms of purified categories. The emphasis on hybrids is not an attempt to invalidate distinctions between entities but rather to deconstruct essentialist categories and build a new world composed of heterogeneous networks, assemblages, or rhizomes.[53]

Meyer argues that landscape architecture should be understood as a hybrid activity that cannot be described using binary pairs as opposing conditions, and instead, should be understood as a combination of human and nonhuman processes. She describes Olmsted's Back Bay Fens in Boston as a "both/and" project, not a recreational facility or a wastewater treatment facility but a multifunctional, multifaceted facility with multiple overlapping meanings: a landscape cyborg, a hybrid of human and nonhuman processes.[54] Hough argues that such landscapes recognize the "interdependence of people and nature in the ecological, economic, and social realities of the city."[55] Appreciating and interpreting such projects requires a perspective that focuses on the relationships between the elements of the project rather than on the elements themselves. In this sense, landscape architecture is not a form of environmental art or a scientific endeavor but a process of relation building, and stormwater management can be understood as a particular configuration of relations among waterflows, humans, technological networks, vegetation, nonhuman species, and so on.

A relational perspective also has important implications for the notion of space. Following on the work of geographer David Harvey, planning theorist Jonathan Murdoch argues that "spatial properties cannot be distinguished from objects 'in' space and space itself can only be understood as a 'system of relations.'"[56] This emphasis on space as a product of relations is in direct opposition to the conventional understanding of space as preceding everything else. Space is constructed within networks, and although it is partly physical, space is wholly relational, presenting a direct challenge to the structuralist geography of well-ordered Euclidean spaces.[57] Latour describes the problem with a topographical interpretation of technological networks as follows: "Between the lines of the network there is, strictly speaking, nothing at all: no train, no telephone, no intake pipe, no television set. Technological networks as the name indicates, are nets thrown over spaces, and they retain only a few scattered elements of those spaces. They are connected lines, not surfaces."[58]

Meyer applies this critique to modernist architects of the twentieth century, such as Le Corbusier, who promoted the juxtaposition of nature and culture and used geometry to perfect and tame a chaotic and undisciplined natural world. She writes, "Architecture is the positive object and nature, the neutral open ground plan patiently awaiting the building's arrival to give it a presence, to give it form."[59] Such a perspective continues to be promoted by planners and architects, and is perhaps most prominent in the New Urbanist discourse that forward geometry as the ordering device for human settlement.[60] Meyer writes, "Hydrological structure, topographic structure, and geological formations are circumstantial obstacles to be overcome or ignored when laying out the rational, universal grid. Landscape does not have another order. It has no order."[61] The modernist perspective of urban planners and architects is also embraced by engineers who shape urban runoff patterns. Their emphasis on network building and technical efficiency encompasses a relational perspective—but one that is abstract or metaphorical rather than grounded in particular places.

Instead of focusing on space as a well-ordered network with absolute and fixed coordinates—*topographic* space—relational theorists argue that we should investigate *topologic* space that emphasizes relations.[62] Space is no longer an empty container waiting to be filled by social or natural activities but rather an active presence held together by stable links or relations.[63] Urban analysis is then the study of associations or networks; the design disciplines are engaged in reconfiguring these associations.[64] An important implication of the topologic understanding of space is that it rejects the possibility of omnipresent, universal rationalities. Whatmore argues that topologic geographies cannot be other than plural and partial, a perspective that is congruent with the messiness of reality.[65] Furthermore, we cannot adopt a vantage point outside of these networks—what Haraway refers to as the "God Trick"—because we are always inextricably bound up in the middle of the landscape and are not afforded a view from the outside.[66] This partial, situated perspective presents a potential source of friction with proponents of ecology who embrace notions of premodern holism and the idea that a single entity comprises all relations. Perhaps the most famous example of this is James Lovelock's Gaia hypothesis, which understands the Earth as an enormous, self-regulating entity.[67] Relational thinkers are skeptical of the common trope of universality because of its tendency to totalize relations and develop yet another grand theory of everything.[68] Rather than moving further and further away from the world to obtain a fuller view, a topologic understanding of space requires urban practitioners to adopt a local perspective to interpret and shape hybrid relations.[69]

An Example of Relational Stormwater Management: Village Homes

Returning to the topic of urban runoff, a relational perspective provides an alternative to the Promethean Project of conventional stormwater management by recognizing the inherent messiness of cities. Rather than attempting to reform conventional stormwater management as prescribed by source control advocates, a relational perspective requires a fundamental reconsideration of the connection between humans and their material surroundings. Practices of stormwater management are recast as processes of relation building that involve technological and ecological knowledge but also cultural and political knowledge. Instead of being a background practice of urban management, stormwater management suddenly becomes an integral part of daily urban life because it is indelibly hitched to all activities.

A small number of urban practitioners have promoted a relational understanding of urban runoff by defining space through the connectedness of landscape and humans. One example of a topologic development with urban runoff as a central element is Village Homes, a sixty-acre master development in Davis, California. Designed and developed in the 1970s by Mike and Judy Corbett, the project consists of 242 clustered houses connected by walking and bicycling paths and a generous amount of open space and community agricultural land (see figure 2.2). The Corbetts combined ideas from Ebenezer Howard's Garden City model at the turn of the twentieth century with the social and environmental movements of 1960s and 1970s; today, the project is celebrated as one of the most successful sustainable community developments in the United States.[70]

The two principal goals of Village Homes were to reduce residential energy consumption and to emphasize community cohesion. The energy conservation emphasis of the project is consistent with contemporary green building practices that use technological and design strategies to reduce the ecological footprint of human habitation. The houses are oriented to the south and are outfitted with photovoltaic solar panels and solar water heaters, resulting in energy savings of about one-third when compared with conventional houses in the region.[71] However, it is the

Figure 2.2
The design of Village Homes includes strategies such as ribbon curbs (left), bioswales and lush landscaping (top right), and the careful siting of housing with respect to existing drainage pathways (bottom right). *Source*: Govind Gopakumar.

second goal of community building that reflects the relational perspective of urban ecologists. The project is most easily traversed on foot or bicycle, with narrow cul-de-sac streets for cars and only 15 percent of the project devoted to streets and parking, as opposed to 22 percent in surrounding neighborhoods.[72] The houses are located on small private lots connected by commonly owned and maintained open spaces as well as community agricultural lands. The Corbetts designed a natural stormwater system of bioswales and ponds in these community spaces to reduce hydrologic impacts and long-term maintenance costs, similar to Low Impact Development strategies.

The topologic character of the development is in direct contrast to designers who advocate for a traditional urban grid pattern. This conflict in design approaches was readily apparent when the Corbetts collaborated with New Urbanist practitioners such as Peter Calthorpe, Andres Duany, and Elizabeth Plater-Zyberk in the early 1990s to develop the Ahwahnee Principles. The merging of New Urbanism, understood as a socially oriented planning logic, and the Corbetts form of ecological planning resulted in principles that had much in common with Ebenezer Howard's Garden City principles, but there was significant disagreement over the layout of streets. The New Urbanists argued for a topographic layout of streets on a strict orthogonal grid, and the Corbetts saw a topologic design conforming to predevelopment site conditions as more effective at addressing both social and environmental flows.[73] The Corbetts pointed to a postoccupancy study of Village Homes that showed residents having twice as many friends and three times as many social contacts as conventional neighborhoods, challenging the New Urbanist argument that ecological design sacrifices social cohesion for environmental protection.[74]

Beyond its spatial layout, Village Homes is intriguing due to the politics involved in the design, construction, and maintenance of the project. Neither of the Corbetts is licensed as an architect, engineer, or landscape architect, but together they have effectively trespassed into the design disciplines to reorient the practice of urban development for their own ends.[75] However, both were dedicated students of landscape architecture, ecology, and site planning and have also been prominent actors in local politics.[76] The design of the project was heavily orchestrated by the Corbetts, but subsequent construction and maintenance has been the province of the homeowners' association with a deliberate emphasis on self-governance and democratic community politics. As such, Village Homes can be understood as a project that challenges modernist dichotomies of public/private, constructed/unconstructed, and social/natural by paying close attention to the intertwined movement of people, water, energy, air, and materials.[77] The design focuses less on the individual elements of the neighborhood and emphasizes the interrelations between the material and social flows of urban life. This approach is not without its faults and can sometimes result in unanticipated problems. For instance, with the blurring of public and private, there have been issues with residents in neighboring communities helping themselves to fruit grown in the community orchards.[78] However, these issues do not overshadow the larger connections that exist between residents and the landscape at Village Homes.

Despite the accolades placed on the project, there has been no replication of Village Homes either in Davis or other U.S. communities. A number of reasons are given for this: a lack of dedicated individuals like the Corbetts to shepherd alternative design strategies through a sea of unsympathetic municipal regulations; a lack of financing opportunities from risk-averse lending institutions; and an American public that does not appreciate community building because of its perceived sacrifice of individual privacy and autonomy.[79] Furthermore, the once-affordable neighborhood of Village

Homes has become a desirable subdivision and is now one of the more expensive places to live in Davis. Social equity has been sacrificed in the pursuit of more environmentally friendly residential living.[80] However, many of the design ideas of Village Homes have been applied to other notable master planned developments, such as Civano (Tucson, Arizona), Prairie Crossing (Grayslake, Illinois), and Coffee Creek (Chesterton, Indiana) in various ways.[81]

Village Homes and its successor projects are important because they demonstrate a gradual reconciliation between the scientific and artistic camps of landscape architecture and ecological planning, as evidenced by a growing number of theorists and practitioners who embrace both the aesthetic and scientific aspects of site design. Although McHarg is frequently criticized for forwarding a rational method for landscape design, he catalyzed a new generation of landscape theorists and practitioners to reassert the importance of the landscape perspective in a variety of ways.[82] His approach to urban runoff has been incorporated and modified by landscape practitioners such as William Wenk, Andropogon, Murase Associates, and Herbert Drieseitl to address both ecological and social flows in urban stormwater projects such as Shop Creek in Denver, Colorado; the Water Pollution Control Laboratory in Portland, Oregon; and Potsdamer Platz in Berlin, Germany.[83] Like Village Homes, these projects interpret cities as relational achievements and suggest how the connections between urban residents and their material surroundings can be configured differently.

Conclusions

Source control as an alternative to the Promethean approach of stormwater management serves as a reform-based movement to significantly change how urban runoff is addressed. The approach also hints at more radical forms of urban water practice as first proposed by Olmsted in the nineteenth century. An influential group of contemporary landscape theorists and practitioners advocate for a very different conception of landscape design, one that does not fall back on the false choice between natural science or aesthetics but rather embraces a relational perspective. Such a logic not only has implications for the technical practices of urban runoff but also more profoundly challenges the idea that cities are devoid of nature or that urban nature is something to be controlled. Ultimately, the challenges to conventional stormwater management demonstrate that the shaping of urban water flows is a deeply political endeavor (I return to the politics of urban drainage in chapter 7). In the following chapters, I use this relational perspective to examine historical and contemporary practices of stormwater management in Austin and Seattle. The struggles with urban runoff in these cities highlight the limitations of attempts to control nature and provide opportunities of conceiving of urban nature in different ways as well as adopting new modes of design practice that can reconfigure cities to reflect their relational status.

3

Saving the Springs: Urban Expansion and Water Quality in Austin

On the evening of June 7, 1990, the Austin City Council held a public meeting to discuss a proposed four-thousand-acre development in the southwest section of the city. A land development corporation was seeking permission from the council to establish a Public Utility District in the Barton Creek watershed that would provide infrastructure services for 2,500 single-family homes, 1,900 apartment units, 3.3 million square feet of commercial space, and 3 golf courses. More than nine hundred citizens attended the public meeting and more than six hundred addressed the city council, the majority of whom were strongly opposed to the proposed development due to its potential impacts on the environment and existing community. The council listened patiently and even extended the meeting to accommodate the record number of citizens who wanted to voice their opinions about the future of Austin. Just before dawn, the council voted unanimously to deny the developer's request.

The notorious All-Night Council Meeting of 1990 was a turning point for the citizens of Austin. The public forum effectively married the community's historic tradition of grassroots community action with its growing concerns for water quality protection, and in the process, changed the character of local politics in the city forever. The meeting would serve as a tipping point between unbridled urban development activities in the 1970s and 1980s to a more managed growth approach in the 1990s and 2000s. In this chapter, I examine the struggles of Austin residents to balance quality of life and environmental protection with economic development and property rights. Local politics in Austin is often characterized by simplistic dichotomies of preservation versus growth, locals versus outsiders, and environmentalists versus developers, but issues of urban runoff transcend these categories to reveal the tensions between culture and nature, built and unbuilt, past and future.

Barton Springs: The "Soul of Austin"

The capital city of Austin, Texas, is located in the middle of the state, framed by the metropolitan triangle of Dallas/Fort Worth, San Antonio, and Houston (see figure 3.1). The city is the fourth largest conurbation in the state, with a population of about 650,000 in the city limits and 1.25 million in the metropolitan statistical area.[1] The location of Austin was selected due to its proximity to water, notably the Colorado River but also a series of creeks that feed into this interregional waterway. A map of the city from its founding year of 1839 situates the 640-acre settlement on the banks of the river where "the new city is shown nestling, as it were, between the sheltering arms of the two creeks, Shoal and Waller" (see figure 3.2).[2] Since the nineteenth century, Austinites and visitors have expounded on the unique beauty of the water features of the city, with the Colorado River and contributing creeks as defining elements of the region.[3]

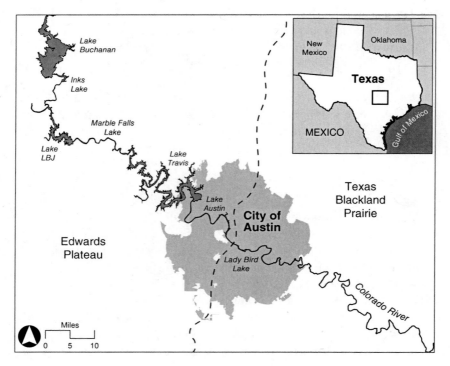

Figure 3.1
Location of Austin on the Colorado River on the cusp of the rolling Hill Country of the Edwards Plateau to the west and the flat coastal plain of the Texas Blackland Prairie to the east.

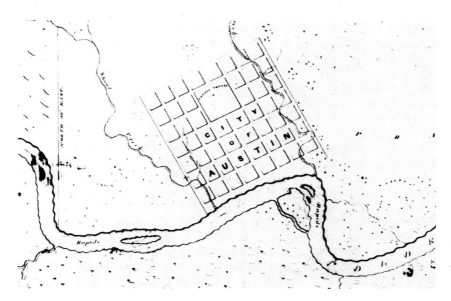

Figure 3.2
A detail of the 1839 Sandusky Map of Austin showing the original city limits bounded by Shoal and Waller Creeks with Barton Creek in the bottom-left corner. *Source*: Map L-2, Austin History Center, Austin Public Library.

The focal point of environmental politics in Austin is not the two creeks that define the original east and west boundaries of the city or the Colorado River but rather Barton Springs, a cluster of four freshwater springs at the mouth of the Barton Creek watershed in the heart of the city. The springs have been a noted locale for thousands of years; archeological and historical evidence suggests that Native American settlements were supplanted in the nineteenth century by Anglo settlers who valued the site as a public meeting place for recreation and community building. An 1884 newspaper article describes the springs as "Austin's Eden," and this sacred interpretation is echoed by a long-time swimmer who states that "Barton Springs is like heaven, more a state of mind than a place."[4]

The municipal government acquired Barton Springs and fifty acres of surrounding land in 1917 from Andrew Zilker, one of Austin's founding fathers, and built a dam below the springs to create a public recreational pool. In 1931, Zilker sold an additional three hundred acres of adjacent land to the municipality, stating that it was a "wrongful thing for this beauty spot to be owned by any individual . . . it ought to belong to all of the people of Austin."[5] In the 1930s, the municipality used funds from

Figure 3.3
Barton Springs Pool today.

the federal Works Progress Administration program to build concrete walls and create the 997-foot-long, three-acre pool that exists today (see figure 3.3).[6] The central location of the pool is inviting to a wide range of Austin residents and visitors, particularly during the hot Texas summers. The springs feed thirty-two million gallons of water per day into the pool at a constant year-round temperature between 68 and 71 degrees Fahrenheit. In the 1920s, the municipality began promoting the springs as a tourist attraction; today, the pool hosts some two hundred and fifty thousand visitors per year.[7]

The springs are near the geographic center of the city and are also central to Austin's culture, providing a direct connection among Austinites, their community, and the landscape.[8] Indeed, the springs are often referred to as the "soul of Austin" and purportedly reflect the egalitarian values held by the city's residents. One longtime swimmer of Barton Springs notes that the springs have "a social leveling influence that is unique. When you strip down to a bathing suit, everyone is equal."[9] Former Texas agriculture commissioner and nationally known political commentator Jim Hightower adds, "The glory of Barton Springs—and similar jewels around the world—is not simply its passive pleasures and the pureness of its existence but also its power to pool individuals into a genuine community of activist citizens. . . . The Springs refresh, but they also empower those who love them."[10]

The springs are understood as the material manifestation of the city's unique culture. When local residents describe the differences between Austin and larger, more politically conservative cities in Texas, the springs are inevitably used as a primary example of how Austin is special.[11] A long-time environmental activist notes, "Houston and Dallas and most communities do not have a focal point to look at and form an approach to managing their future. But we have Barton Springs."[12] In the 1970s, a popular local campaign called for residents to "Keep Austin Austin," and in the last decade, the slogan "Keep Austin Weird" has come to define a shared desire to retain the unique characteristics of the city.[13] A local journalist summarizes the connection between Austin's identity and the springs, stating, "The idea that Barton Springs and Austin are one and the same, that the Springs do incarnate what separates Austin from Everywhere Else, and that living in Austin carries with it an obligation to protect Barton Springs from harm."[14]

Beyond its uniqueness, the community often romanticizes Barton Springs as a reflection of past conditions of the city, when the life in Austin was simpler and was defined by more leisure and communal activity and less work. The photographic record of the springs from the 1920s to the present suggests that little has changed there in the intervening years; the pool has been frozen in time as a place for relaxation, fun, and community building. An environmental activist conflates the history of the springs with the character of Austin and its residents, noting that the springs are "a symbol of our connection to something really ancient and old in terms of this amazing, clean, clear water supply, the history of the Springs. And for people who never swim . . . even for people who never dip their toe in that water, it stands for a connection to our history in this town, and everything that is special and unique about Austin."[15]

The springs offer respite from the hustle and bustle of the increasingly urbanized surroundings as well as a direct connection between Austinites and nature. A brochure from the municipality's Parks and Recreation Department describes the springs as "an island of nature in an ocean of urban development" and "a natural refuge so close to the heart of downtown."[16] This perspective reinforces the common understanding of cities in opposition to nature. However, upon closer examination, it is clear that the interpretation of the springs as "pure nature" is disingenuous. The highly constructed elements of the pool include concrete sidewalls, modern

Figure 3.4
A 1948 photo of Barton Springs Pool showing construction of a conduit to channel Barton Creek around the pool. *Source*: PICA 20225, Austin History Center, Austin Public Library.

changing facilities, a security fence, and of course, the dam that creates the pool of water.[17] Historic photos reveal the technically mediated nature of the pool, as does the monthly cleaning process that requires the partial draining of the pool (see figure 3.4). Though having less rhetorical power, the springs and pool should be more accurately described as a hybrid of natural, technical, and cultural processes that forge a unique relationship between Austinites and their material surroundings.

Situated just upstream from the Barton Springs Pool is the 7.9-mile, 809-acre Barton Creek Greenbelt, a recreation destination for hikers, mountain bikers, rock climbers, and swimmers. The greenbelt, completed in the late 1980s, extends the notion of nature in the city with a cobble-bedded creek, a dense tree canopy, sheer rock walls, and a primitive trail system.[18] The greenbelt and springs reflect the "green romanticism" embraced by many Austinites who forward the identity of the community as rooted strongly in a pure, virgin nature.[19] There is a strong sense of place informed by a reverence for nature and a fetishization of these places as the genuine character of the city. However, the importance of this perspective becomes clear only when considered with respect to upstream land development activities.

Harnessing the Colorado River

The cultural importance of the springs to Austin is equaled by the economic importance of the Colorado River and its promise as a consistent source of water and electricity. The river is nearly six hundred miles long, comprising 39,900 square miles of catchment from West Texas to the Gulf of Mexico.[20] Visitors from other regions often expect to find a dry landscape devoid of vegetation—a stereotype of Texas portrayed in Western movies—but the environmental conditions in Austin are strikingly different. The city receives a moderate thirty-two inches of rain per year on average and the region is subjected to some of the largest flood-producing storms in the continental United States.[21] It is a region with an abundance of water, although it tends to come in short bursts rather than in regular rainstorms throughout the year. Reflecting on the flooding threats in the city, an Austin historian notes that "if Noah had lived in Austin he would have been much more easily persuaded to build the Ark."[22]

The challenges of settling in such a flood-prone area were evident in the late nineteenth and early twentieth centuries as the municipality struggled to protect its citizens and property from frequent and devastating storm events. The City of Austin built its first dam on the Colorado River in 1893 to generate electricity and provide flood control as well as a consistent source of drinking water, but the river resisted human manipulation and washed the dam away only seven years after completion.[23] The failure of the dam sent the city into fiscal crisis and created an unstable environment for economic investment; between 1900 and 1913, Austin residents suffered through seventeen floods that caused $61.4 million in damage and claimed the lives of sixty-one residents (see figure 3.5).[24]

The threat of nature galvanized Austinites to adopt a Promethean approach to tame the Colorado River and create a safe and stable foundation for urban development. Austin was more fortunate than other cities in Texas, receiving the highest amount of funding from the Public Works Administration in the 1930s for municipal construction projects.[25] Lyndon Baines Johnson, then a U.S. congressman, and other politicians were instrumental in bringing New Deal projects to Central Texas, particularly the six dams on the Colorado River that created the hundred-and-fifty-mile chain of water bodies known as the Highland Lakes (see table 3.1

Picturesque and peaceful Barton Springs was turned into a raging torrent that left debris and destruction after this flood rolled in the morning of June 17, 1958.

Figure 3.5
The 1958 flood at Barton Springs Pool illustrates the destructive potential of regional hydrologic events. *Source*: PICA 22648, Austin History Center, Austin Public Library.

Table 3.1
The Highland Lakes system

Dam	Date of completion	Body of water created	Surface area (square miles)
Buchanan Dam	1937	Lake Buchanan	36.03
Inks Dam	1938	Inks Lake	1.25
Tom Miller Dam	1940	Lake Austin	2.86
Mansfield Dam	1941	Lake Travis	29.58
Starcke Dam	1950	Marble Falls Lake	1.22
Wirtz Dam	1950	Lake LBJ	9.96

Source: Banks and Babcock 1988.

and figure 3.1). The dam projects created jobs and provided residents in the region with much needed water and electricity services as well as flood control.[26] Austin historian Anthony Orum notes: "Harboring few natural resources that could be mined for their wealth and usefulness, set in a wooded, hilly landscape, bedecked with rivers and lakes, Austin was a prisoner of its environment, yet one that enjoyed its fate. But the dams had changed all that. They had made possible a control of the rivers and industry that could use the electric power they harnessed. They had made life livable in Austin, and throughout the rest of Central Texas."[27]

Reflecting on the damming of the Colorado River, Johnson stated in 1958:

Of all the endeavors on which I have worked in public life, I am proudest of the accomplishment in developing the Colorado River. It is not the damming of the streams or the harnessing of the floods in which I take pride, but rather in the ending of the waste of the region. The region—so unproductive in my youth—is now a vital part of the national economy and potential. More important, the wastage of human resources in the whole region has been reduced. Men and women have been released from the waste of drudgery and toil against the unyielding rocks of the Texas hills. This is the true fulfillment of the true responsibility of government.[28]

Johnson's emphasis on "waste" and "productivity" reflects the central themes of the Progressive era at the turn of the twentieth century as well as New Deal politics in the 1930s and 1940s. Undoubtedly, he was a great believer in the Promethean Project as a means to fulfill the progressive goal of a more prosperous future.

Once the dams were in place, the latent potential of Austin as an attractive Sunbelt city could be realized by the municipal government, the Chamber of Commerce, and other regional development interests.[29] The city now had a stable foundation on which to expand its economic base beyond its original economic pillars of state government and higher education to include light manufacturing and, more important, the high-tech sector. The region quickly received national attention for its economic development potential, as evidenced in a 1950 *New York Herald Tribune* article: "A sizeable block of the people of the United States (estimated conservatively up in the millions) today are conscious of the fact that there is such a place as Austin, Texas, and that at and near Austin are a series of fresh water lakes nestling in the picturesque hills, which not only provide opportunities for outstanding recreation, but also contain ample water for development of cities, agriculture and commercial and industrial projects."[30]

With stable municipal services and a burgeoning recreation economy from the Highland Lakes, the Austin population surged. From 1940 to

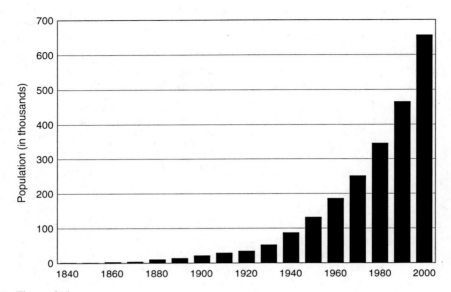

Figure 3.6
The exponential population growth of Austin from 1840 to 2000. *Source*: City of Austin 2007a.

2000, the Austin population grew at an average rate of 40 percent per decade due to aggressive marketing by a consortium of university administrators, municipal officials, and local entrepreneurs (see figure 3.6).[31] IBM was the first major high-tech company to move to Austin in 1967, followed by Texas Instruments in 1969, Motorola in 1974, and numerous others in subsequent decades. In 1988, *Inc.* magazine named Austin the leading entrepreneurial city in the United States due to the large number of new enterprises being created.[32] Companies were attracted by Austin's affordable land, an educated and low-wage workforce, a university with strong science and technology programs, access to growing southwestern markets, and—perhaps most important—an attractive environment for residential life.[33] The social and material conditions created a winning formula for realizing "the good life" both at work and at play by embracing an "urban growth machine" strategy.[34]

The urban growth machine narrative of Austin's development diverges sharply from the green romanticism of the residents at Barton Springs. The dissonance between these two interpretations of Austin's future would serve as the crux of local political conflict beginning in the 1970s that pitted growth proponents against environmental preservationists. One side perceived the landscape as a bundle of resources to be harnessed for human ends; the other saw the landscape as integral to the unique culture of Austin and in need of protection from future development. It was an urban manifestation of the classic U.S. environmental debate between conservationists and preservationists in the late nineteenth and early twentieth centuries.[35] Nature could be revered or it could be exploited—but not both.

Urban Growth Politics in the 1970s

The urban growth that resulted from the booming high-tech industry was a boon to landowners and developers but was increasingly unwelcome to many Austinites who perceived the rapid population increase and land development activities as a threat to their cherished quality of life. Austin's politically liberal population, based on employment in state government and higher education, created a community of highly educated middle-class residents. The University of Texas fueled the activist character of the community beginning in the 1930s, when university professors and students railed against the Austin establishment on contentious social issues, notably segregation.[36]

In the 1970s, neighborhood groups began to focus on activities to slow or stop the explosive urban growth patterns fueled by aggressive economic development activities. Neighborhood activists fought battles over new apartment complexes and traffic congestion to protect the integrity of their neighborhoods from new development; by 1983, there were more than a hundred and fifty neighborhood groups active in Austin as well as numerous local environmental organizations that focused on protecting undeveloped land from environmental degradation.[37] The environmental and neighborhood groups often joined under a shared banner: "limit growth, limit urban expansion."[38] As such, it was in the 1970s that environmental protection became synonymous with protection of Austin's cherished quality of life.[39] The green romantics in Austin questioned the assumption that urban growth was an inevitable and positive outcome of societal progress and instead promoted an idea of bioregionalism and respect for place.[40] This perspective is reflected in the militant, uncompromising position of famed Texas naturalist Roy Bedichek:

Personally, if I have to fight for this country, I will not fight for the flag, or the American "way of life," or democracy, or private enterprise or for any other abstractions, which seem cold as kraut to me. But I will fight to the last ditch for Barton Creek, Boggy Creek, cedar-covered limestone hills, Blazing Star and

Bluebonnets, Golden-cheeked Warblers and Black-capped Vireos, and so on through a catalogue of this natural environment of Austin, Texas. It is through this natural environment of Austin, Texas, that I love America.[41]

A Fragile Land: The Texas Hill Country

A particular focus of Austin residents, regional politics, and urban growth is the land to the west of the city known as the Hill Country. The boundaries of the Hill Country are loosely defined but comprise an area spanning several counties and hundreds of square miles of largely rural area characterized by rolling landscape, massive live oak trees, spectacular fields of wildflowers, and picturesque views.[42] There are a plethora of historical comparisons between the Hill Country and Mediterranean environments, particularly the Tuscany region of Italy.[43] However, the appearance of a bountiful landscape that could rival the agricultural production of Tuscany was deceptive. Famed LBJ biographer Robert Caro characterizes the scenic beauty of the Hill Country as a trap for early settlers:

[The grass had] grown so slowly because the soil beneath it was so thin. The Hill Country was limestone country, and while the mineral richness of limestone makes the soil produced by its crumbling very fertile, the hardness of limestone makes it produce that soil slowly. There was only a narrow, thin layer of soil atop the Hill Country limestone, a layer as fragile as it was fertile, vulnerable to wind and rain—and especially vulnerable because it lay not on level ground but on hillsides: rain running down hillsides washes the soil on those slopes away. The very hills that made the Hill Country so picturesque also made it a country in which it was difficult for soil to hold. The grass of the Hill Country, then, was rich only because it had had centuries in which to build up, centuries in which nothing had disturbed it. It was rich only because it was virgin. And it could only be virgin once.[44]

It was this virginal appearance of the Hill Country and its promise for great agricultural productivity that would attract earlier settlers entranced by the Edenic character of the landscape.[45]

Early land use in the Hill Country consisted largely of cattle ranching and cotton farming, activities that had devastating effects on the soils and vegetation by compacting the land and encouraging runoff during storm events. Although agricultural land use is different from urbanization, it has similar physical effects in terms of producing polluted storm-water runoff. These impacts were amplified as land use in the Hill Country gradually expanded to include suburban residential and commercial development. Today, the area continues to support cattle ranching but the region as a whole is increasingly characterized by single-family residences in suburban subdivisions.[46]

Going Underground

The land development activities in the Hill Country would affect not only surface features but also underground water flows. The subsurface of the region is a vast underground storage vessel of permeable limestone characterized by caves, sinkholes, and other conduits. Due to its porous character, the Edwards Aquifer is the most environmentally sensitive subsurface waterbody in Texas, in direct contrast to most aquifers that filter groundwater. The porosity of the aquifer results in direct interaction between activities on the land and conditions of water quality below ground. Soil erosion in the Hill Country has an impact on not only the immediate, visible landscape but also the water quality of the aquifer and receiving waters—notably Barton Springs. It is this sensitive underground network of water flows that connects land development in the upstream Hill Country to Austin culture downstream.

Formal studies of the aquifer's complex flow paths began over a century ago with geologic mappings and groundwater monitoring. In the 1950s, University of Texas geologists began large-scale studies of the Edwards Aquifer; today, they continue to refine their understanding of aquifer mechanics using increasingly sophisticated field studies and hydrologic flow models. In 1978, the City of Austin initiated a partnership with the U.S. Geological Survey to conduct comprehensive studies to characterize aquifer flow; in subsequent years, they developed large datasets, models, and theories to delineate how the aquifer functions. A municipal staff member notes, "When I first started here in the mid 1980s, we were assuming that it might take twenty years for water to flow through the aquifer. Now, we know that it is a matter of days or even hours. So the leap in knowledge has been pretty amazing."[47]

Despite the intensive geologic research projects by university, municipal, and federal entities, the diverse flow paths and unexpected behavior of contaminants in the aquifer continue to confound scientists. For example, the results of dye tracer studies completed by the City of Austin and the regional aquifer authority in the late 1990s negated some of the earlier understandings of the aquifer's flow paths. The aquifer acts as a "trickster," eluding precise scientific prediction and fostering uncertainty and doubt with its complex behavior.[48] In 1997, a consortium of scientists, engineers, and planners published a scientific consensus document in an attempt to solidify the commonly understood characteristics of the aquifer, but the uncertainty of land use development impacts on the aquifer continues to hamper efforts to balance urban growth with environmental protection.[49]

The aquifer is divided into three parts: the Northern Edwards section, the Southern Edwards section, and the Barton Springs section. The latter section, often referred to as the Barton Springs Zone, is 355 square miles of land straddling two counties (Travis and Hays) and comprising six creek watersheds: Barton, Williamson, Slaughter, Bear, Little Bear, and Onion.[50] As its name implies, the Barton Springs Zone is a hydrogeologic funnel that discharges at Barton Springs, and it is through this direct connection to the Edwards Aquifer that the Hill Country landscape takes on cultural importance to Austinites. As development occurred in the region, land use practices produced urban runoff that introduced contaminants into the aquifer, and these contaminants traveled quickly and largely unfiltered until they surfaced at Barton Springs. Thus, in addition to being a culturally relevant locale, the pool became an indicator of regional environmental health, serving as a monitor for the environmental impacts of land development activity upstream of the springs.[51]

The increasing recognition of the environmental sensitivity of Barton Springs to upstream development activities was not lost on Austinites and municipal officials. In the 1970s, growth issues in the Hill Country quickly dominated local politics as residents feared the negative environmental and cultural impacts of upstream development on the springs.[52] An Austin commentator, in an opinion piece titled "The Screwing Up of Austin," sums up the 1970s attitude of local residents: "Pick up any outlander newspaper or magazine these days, and you are liable to read about the peculiar appeal of Austin. They're catching on to it out there, partly because of these honkytonk heroes and motion picture gypsies who are slipping in, and partly because a number of greedheads who discovered they could make a living out of Austin by chopping it down."[53]

One of the most significant efforts by the municipal government and citizens to strike a balance between urban development, environmental protection, and quality of life was the *Austin Tomorrow Plan*, a large-scale comprehensive planning effort initiated in 1974. Through a participatory process that involved thousands of citizens as well as numerous neighborhood groups, business interests, and municipal staff members, the municipality devised a comprehensive plan that was eventually adopted as official municipal policy by the city council in 1980.[54] A significant element of the plan called for limited development in environmentally sensitive areas including the Barton Springs Zone and in areas next to waterways. However, the desire to protect the aquifer and to steer growth to other parts of the city, as outlined in the comprehensive plan, was not realized. In the late 1970s and early 1980s, voters failed to ratify several water and wastewater bond referenda that would have provided necessary facilities and improvements to create the desired north–south growth corridor as specified in the *Austin Tomorrow Plan*.[55] A municipal staff member notes, "When I think about the early master plans that so clearly said that we don't want development to the west, there were no mechanisms to make that happen. It was just market forces; people want to be in the Hill Country."[56]

The Hill Country was attractive because of its close proximity to Austin's burgeoning employment opportunities, its natural beauty, and an abundant supply of water from the Highland Lakes, the most crucial element to urban development in the region. As the high-tech sector fueled population growth, developers initiated large-scale suburban development in the Hill Country, in direct contradiction with the *Austin Tomorrow Plan*. The municipal government's planning commission and city council were complicit in these development projects, frequently offering exemptions to subdivision and land use regulations to allow new development over the aquifer, consistent with the municipality's historic emphasis on economic development.[57]

Development in the Hill Country was not only encouraged by the municipal government but also by State law that allowed developers to create Municipal Utility Districts (MUDs). MUDs were enacted by the state legislature in 1971 as a mechanism to finance water, wastewater, and drainage improvements for suburban developments while superseding local regulations. Today, the greatest concentration of MUDs can be found in the sprawling metropolitan area surrounding Houston—a region notorious for its lack of an urban growth strategy—but Austin also experienced significant MUD development in the 1970s and 1980s.[58] MUDs effectively uncouple infrastructure provision from municipal governance and allowed developers to provide their own essential services.[59] Urban expansion then becomes the providence of market forces rather than comprehensive municipal or regional planning, a defining characteristic of Texas land development activities.

In addition to MUD developments, infrastructure services are increasingly provided by the Lower Colorado River Authority (LCRA), the state-chartered entity that manages the Highland Lakes. The LCRA is purportedly an apolitical organization, but its general manager is chosen by a board of directors appointed by the governor of Texas and its income is derived from water and electricity sales. As such, the LCRA is often characterized as a promoter of land development and an alternative infrastructure provider to the City of Austin. The LCRA's former general manager defends the growth-friendly position of his organization: "If we have the water, then legally we have to provide the water."[60] Urban theorist Steven Moore links the water provision mandate of the LCRA to its larger role in regional development: "This remarkably important institution effectively controls water policy in a semi-arid region, and thus by default it controls rural development policy, with neither accountability to citizens nor any responsibility to coordinate their policies with regional city governments."[61]

The presence of the LCRA expands infrastructure provision beyond the exclusive, monopolistic practices of local government to include state and private interests and offers multiple opportunities to provide essential services for new land development projects. This mirrors the trend of "splintering infrastructure" that emerged in the 1980s due to privatization and decentralization; service provision is no longer tied to the monopoly of the municipal service provider but is instead guided by free market principles of economic liberalism.[62] As such, negotiations over infrastructure provision in the Hill Country have become the primary means of managing growth, supplanting zoning and subdivision regulation.[63]

Reflecting on infrastructure service provision in the Hill Country, an LCRA engineer notes, "We frame [water quality protection] as an option as opposed to a rule. . . . We also try to be a friendly regulator . . . we try to help people with technical assistance." The respondent goes on to describe a technomanagerial approach to infrastructure provision and environmental quality that emphasizes sound technical and scientific decision making that is purportedly free of troublesome growth politics: "LCRA is much less political than the City of Austin. It's so much easier to do our job here. We do what we think is right based on the technical findings and the science and not have all of the politics weighing in on something entirely different that might direct the policy of the City of Austin."[64] This apolitical approach to urban development and infrastructure provision in Austin resonates with the Progressive era of infrastructure development, with experts tasked with solving a predefined problem: how to provide collective services to a growing populous. But the public discourse in Austin and the Hill Country reveals infrastructure provision and land use development as inherently political activities, ones that grapple with the tensions of living in a fragile landscape. A consulting engineer familiar with the organization provides a different perspective, noting that the "LCRA has strongly resisted the regulatory role; they don't want to be seen as a policeman, they want to be a provider of electricity and water."[65]

The splintering of infrastructure services is exacerbated by Austin's location in the state of Texas, where "home rule" or local control is the law of the land. The state legislature has a long tradition of encouraging municipalities to govern themselves, rather than promulgating state legislation to direct growth, as illustrated by George W. Bush's gubernatorial election campaign theme of 1994: "Let Texans Run Texas."[66] The emphasis on local control results in fragmented land management rather than structures for regional cooperation and coordination for how and where growth should occur.[67] More important than the emphasis on local control in Texas cities is the lack of authority that larger governmental bodies, particularly counties, have over development processes outside of city boundaries. Instead, Texas law allows for "extraterritorial jurisdictions" (ETJs) that enable municipal governments to exercise subdivision and infrastructure regulation up to five miles beyond their corporate limits, depending on the population of the municipality.[68] However, ETJ regulations do not include zoning requirements, resulting in an "anything goes" form of land development in the geographic areas between municipalities.

As a whole, Texas land use policies are the quintessential recipe for urban sprawl: a confluence of strong property rights supporters, weak local and regional governments, and a lack of county and state regulations to direct urban growth. A local journalist attributes the problem of urban sprawl in Texas to a continual denial by state legislators of the importance of large cities in Texas. He writes, "Texas' self-image as a primarily rural, resource-based state, where the cities simply exist as service centers for ranches and oilfields, is taking a long time to die."[69] The frontier mentality of a limitless landscape persists in the mind of Texans, despite the fact that the state is home to three of the ten largest cities in the United States (Houston, Dallas, and San Antonio) and six of the largest thirty cities.[70] These urban areas play an increasingly significant role in regional issues, particularly land use, transportation, water supply, and environmental protection.[71]

Controlling Growth with Water Quality Regulations

The regulatory attitude of the LCRA and the State of Texas is often contrasted with that of the City of Austin and is frequently characterized as facilitating rather than regulating urban development and growth. The LCRA and the state government follow an economic rationalist approach that ties societal progress to property development and sees increasingly dense human settlement of the Hill Country as not only desirable but

inevitable. In contrast, the City of Austin is often characterized as a jurisdiction with a strong commitment to environmental protection and has
passed a number of stringent regulatory requirements to fulfill these goals.
An environmental activist notes:

The City of Austin is the only bureaucracy I have ever seen that is very committed to water quality protection as a community. . . . I think they see themselves as
responsible to the community and responsible for environmental protection in a
way that is pretty unique. I have never gotten that sense from working with the
county, LCRA, or TCEQ [Texas Commission on Environmental Quality, the state
environmental agency], a sense for accountability to environmental protection.
They just don't. What they have a sense of accountability to is the letter of the
law, if that. They are very different cultures.[72]

This emphasis on environmental protection by the City of Austin is another characteristic that sets it apart from neighboring jurisdictions and
the state government.

The municipality's commitment to protecting water quality is most
evident in the series of watershed ordinances passed between 1980 and
1991 that placed limits on development in areas surrounding creeks and
watersheds through the Land Development Code—specifically, subdivision and site planning activities.[73] The ordinances were inspired by the
Lake Austin Growth Management Plan of 1976—one of the earliest examples of water quality planning in the United States—prepared for the
City of Austin by the renowned landscape architecture firm of Wallace,
McHarg, Roberts & Todd.[74] The plan was one of the earliest to develop
a solution to the water quality impacts of urbanization and was based on
scientific correlations between impervious cover, distance to waterways,
and impacts on water quality. Measures included development density
limits and construction bans on steep slopes and near watercourses and
sensitive environmental features.[75]

The first major watershed ordinance to address the impacts of urbanization was the Lake Austin Watershed Ordinance, passed in 1980. The ordinance included impervious cover limits, structural controls, an erosion/
sedimentation control plan, and prohibitions on building in the hundred-
year floodplain around Lake Austin (one of the Highland Lakes). Two
additional watershed ordinances were passed the same year and extended
water quality protection measures to two West Austin watersheds, Barton
and Williamson. In subsequent years, more rural watersheds were added,
as were buffer zones on creeks and sensitive environmental features, and
impervious cover limits were increased. Of particular note was the 1986
Comprehensive Watersheds Ordinance that extended environmental

protection to watersheds that did not provide drinking water for Austin residents. This demonstrated that water quality protection was not only necessary for human health but was also important for nonhumans. The 1991 Urban Watersheds Ordinance required water quality control structures on new development in *urban* watersheds, recognizing that built-up areas also needed to protect natural resources.[76]

The City of Austin's water quality ordinances formalized the municipality's commitment to protecting the region's natural resources and cemented its reputation as a national leader in environmental protection.[77] Moore refers to this period in Austin's history as an era of code building that transformed the environmental values of residents into local government policy.[78] The watershed ordinances are a form of administrative rationalism that incorporate the evolving attitudes of residents toward environmental protection, codifying a new interpretation of how nature and society should relate by translating the latest scientific data to land use regulations.[79]

Sending Out an SOS

The aforementioned watershed ordinances were in effect when the All-Night Council Meeting was held on June 7, 1990, but environmental and community groups felt that these measures were insufficient to protect Barton Springs from future water quality impairment. Citizen groups were incensed by the "grandfathering" of less stringent water quality regulations as well as back-room deal making and variances by the development-friendly city council and planning commission that effectively made the watershed ordinances as porous as the aquifer they were designed to protect. In other words, though the ordinances were on the books, their existence did not always translate into rigorous enforcement. Citizens used the All-Night Council Meeting to persuade the municipal government to take the issue of water quality protection seriously. Riding on the momentum of the All-Night Meeting, a number of local environmental activists formed the Save Our Springs Coalition (later renamed the Save Our Springs Alliance or, hereafter, SOS Alliance) to focus specifically on protecting Barton Springs and all of the watersheds that contribute to the Barton Springs segment of the Edwards Aquifer.[80]

The SOS Alliance quickly drafted a new water quality ordinance that would codify the community's desire to protect the Barton Springs section of the Edwards Aquifer. The activists wanted to institute stringent municipal regulations to permanently prohibit large, unwanted development over the areas of the aquifer in the City of Austin's jurisdiction. The organization gathered over thirty thousand petition signatures to qualify the citizen initiative for the 1992 municipal election ballot, and after contentious political debates and procedural delays by the development-friendly city council, the ordinance was voted into law by a decisive two-to-one margin. According to an environmental activist, the passage of the SOS Ordinance resulted in the "institutionalization" of water quality as the primary agenda of Austin growth politics.[81] Another environmental activist states that "with the SOS vote, Austin made it clear that Barton Springs is where we draw the line."[82]

The SOS Ordinance was different from preceding water quality regulations because it became law by citizen initiative. It is an example of an active citizenry collaborating with environmental experts to push through regulations that municipal officials were unwilling or unable to pass through conventional code-building processes. Specifically, the SOS Alliance challenged the primacy of the municipal government and development community in setting the standards for water quality protection and urban growth. The group successfully transformed the grassroots momentum of the All-Night Meeting into rational government policy.

The water quality protection strategies of the SOS Ordinance were not new, but they were significantly more stringent than the previous watershed ordinances; most important, the ordinance prohibited the granting of variances by the city council and planning commission. The ordinance restricted impervious cover in the Barton Springs Zone to 15 to 25 percent depending on the degree of connectivity between the landscape and Barton Springs. This is roughly the equivalent to half-acre to one-acre residential lot development density. In addition to the impervious cover limits, the ordinance included a nondegradation policy that required technical strategies or best management practices (BMPs) to ensure that the average annual loadings of common stormwater pollutants (suspended solids, nutrients, heavy metals, and so on) did not increase between pre- and postdevelopment conditions. This portion of the ordinance is often referred to as the "nondegradation clause" and essentially requires that new developments do not affect water quality. Finally, the ordinance required building setbacks of two hundred feet from major waterways and four hundred feet from the main channel of Barton Creek. Reflecting on the combination of impervious cover limits and the nondegradation requirement, a consulting engineer states, "There was a fear of technology and the idea that technology was going to fail you. So even though we had these requirements for controls in the SOS Ordinance, we were afraid that they were never going to be maintained and they wouldn't work. So that's why you do impervious cover limits on top of nondegradation."[83]

Impervious cover limits are an easily understandable regulatory requirement that ensure equitable application of environmental protection to all properties in the Barton Springs Zone, but they are also scientifically controversial. The impervious cover limits were reportedly derived from environmental thresholds established by Tom Schueler at the Center for Watershed Protection in Ellicott City, Maryland. Schueler, an early proponent of Low Impact Development and source control methods for stormwater management, developed these thresholds from research data on urbanizing creeks in the Pacific Northwest. The authors of the SOS Ordinance (environmental lawyers, scientists, and engineers) used Schueler's thresholds as scientific justification for water quality, despite the completely different hydrologic and geologic conditions of the Hill Country. Thus, seemingly universal scientific findings were applied to particular local conditions. Moreover, some scientists question the ability of the impervious cover thresholds to reflect the complexities of stormwater runoff and aquifer flow patterns. A local geologist notes:

> The public is only smart enough to understand the link between impervious cover and water quality but this approach is not nuanced enough to reflect how the aquifer functions. We could make a substantive impact by protecting the aquifer's most vulnerable features but this doesn't fit with the legal system where property rights are dispersed. If we could do a regional survey and protect the most sensitive features, we could do a lot more than impervious cover limits. This would allow for development while also protecting the aquifer.[84]

In other words, the problem with impervious cover limits is that the approach treats all properties equally, but the landscape is heterogeneous and includes varying areas of sensitivity that are not evenly distributed.

The parcel-by-parcel application of land use restrictions is a topographic approach to environmental regulation and is in direct conflict with the complex topologic relations of the Hill Country landscape and the underlying Edwards Aquifer. There is a fundamental mismatch between the social system of segmented property ownership and the physical system of landscape connectivity. Reflecting on this clash, an environmental activist involved in the drafting of the SOS Ordinance agrees that "it really is fairer to property owners when it's not as scientifically exact or nuanced."[85] The impervious cover limits effectively call for large-lot residential development on upland surfaces and moderate hill slopes, but it is unclear if this type of development, even if it strictly adheres to the SOS Ordinance requirements, will protect downstream water quality.[86]

Furthermore, the SOS Ordinance is an implicit mandate to limit urbanization. Although the ordinance does not prohibit specific forms of land use, it restricts property development, effectively creating an urban growth boundary between high-intensity urban development and low-intensity rural development. There is a shared concern among the SOS Ordinance supporters and critics that the resulting low density will encourage sprawling suburban development. A consulting engineer notes: "One of the ironic things is that we want to protect water quality by preserving open space but we contribute to sprawl with the impervious cover limits. From an urban planning perspective, I don't think you can build a city with 15 percent impervious cover. What would San Francisco or New York look like? It doesn't make any sense. Mass transit doesn't work, nothing works. Suburbs and Walmarts."[87]

Not surprisingly, the development community and property owners in the Barton Springs Zone who were following the economic rationalist approach of conventional land development perceived the SOS Ordinance as an infringement on their property rights and filed lawsuits against the municipal government. When these preliminary suits were dismissed, the developers successfully lobbied State legislators to pass legislation specifically aimed at weakening the SOS Ordinance (referred to by locals as "Austin-bashing" legislation).[88] The result has been the grandfathering of properties to allow development under previous, less stringent development regulations. However, the May 1998 ruling by the Texas Supreme Court proved to be the final legal word on the SOS Ordinance, stating that "such limitations are a nationally recognized method of preserving water quality [and] the City has the right to significantly limit development in watershed areas in furtherance of this interest."[89]

The SOS Ordinance can be understood as the transformation of green romanticism and cultural identification with the springs into administrative rationality using the most current scientific understandings of urban runoff science and aquifer mechanics. However, conflict arose over the issue of property ownership and the ability of the municipality to redefine property rights to reflect changing social values. Environmentalists interpreted property rights as communal, encompassing social responsibilities and commitments as well as individual privileges; thus, protection of Barton Springs is a collective responsibility to be shared by all property owners. Meanwhile, property rights advocates interpreted the changes to their legal entitlements to be in direct conflict with individualist American values, where it is understood that "to be an American [is] to own and control private property."[90] This strong property rights aura is prevalent in Texas, as an environmental activist notes: "Texas has this love affair with sort of a cowboy image. You get to do what you want to do on your own land and you have a right to do that. And this worked very well when you wanted to put in a windmill pump and run some cattle. And we have carried forward that mentality to a time when what you do on your land makes a very big difference to everybody else. And it just doesn't work very well."[91]

The regulatory approach of the SOS Ordinance extends Austin's cultural interpretation of the springs upstream into the Hill Country. The regulation is a means of translating new understandings of aquifer science to the property ownership regime. However, its translation into legally binding regulation using impervious cover restrictions, treatment strategies, and setbacks is merely a proxy for the ecological flows of the landscape and falls short of reflecting the interconnectedness of regional water flows. In other words, the topology of environmental flows is in conflict with the topographic system of property rights that demarcates the landscape in discrete, bounded units. Despite the passage of the environmental regulations, the aggregate effects of development in the region, even under the stringent limits of the SOS Ordinance, continue to be unknown. The landscape and subsurface continue to resist enrollment in the framework of land use regulation and environmental protection.

On May 30, 1997, the aims of Austin's environmental community were bolstered when the U.S. Fish and Wildlife Service granted the Barton Springs Salamander (*Eurycea sosorum*) protection under the federal Endangered Species Act.[92] The salamander was discovered in the 1940s and is only known to reside in the Springs. Unlike other salamanders, the Barton Springs Salamander retains external gills throughout its life and is dependent on a continuous flow of clean, clear, and cool water.[93] The federal government identified the primary threats to the species as the degradation of water quality and quantity in Barton Springs resulting from urban expansion in the Barton Springs watershed.[94] As such, the role of Barton Springs as an indicator of the ecological health of the aquifer was further solidified, with the salamander serving as the proverbial "canary in the coal mine."[95] Where the development community reached out to the state government to forward their interests, the environmental community trumped this action by invoking the hammer of federal regulation. Protection of Barton Springs was now simultaneously a local issue of cultural preservation and a national issue of biodiversity protection.

Defending the Community's Environmental Interests

Environmental protection in the Hill Country was promoted and defended not only by legal staff at City of Austin, who spent millions of taxpayer dollars to defend the SOS Ordinance against property owners, developers, and the State of Texas, but also the SOS Alliance, an organization that has persevered as a staunch opponent to development and property rights interests. The SOS Alliance initiated its hardliner approach to environmental protection during the SOS Ordinance election campaign of 1992 and continued with subsequent litigation against the municipal government and developers to enforce and uphold the legislation. The SOS Alliance is often compared to radical, ecocentric environmental organizations such as Earth First!, but their tactics are more in line with civil society groups such as Environmental Defense and the Natural Resources Defense Council; it is a confrontational environmental organization that pursues its agenda in formal legal and political arenas.[96]

The environment is framed by the SOS Alliance in preservationist terms as a pristine landscape to be spared from the wanton destruction of profit-driven land speculators and property developers. This is illustrated by a common mantra of the organization: "Save it or pave it." The stark, dualist perspective on the landscape is consistent with the organization's uncompromising, litigious approach that proved highly successful in the early 1990s. However, by the end of the decade, many of its earliest and strongest supporters had moved on, stating that they were tired of the hard-line approach and the vitriolic political atmosphere it created.[97] A former executive director of the SOS Alliance notes, "They play only one way. They're not a finesse organization; they're a fight-to-the-death organization."[98] As such, the politics of environmental protection in Austin, particularly in the 1990s, can be characterized by a highly partisan attitude of either saving Barton Springs or being complicit in its destruction. The environmental politics in the city are restricted to the status of the springs and the upstream land development activities.

Over time, the militant approach has compromised the credibility of the organization and its representation of Austin residents on environmental issues. A former city council member with green credentials argues that "fundamentally, what happened with [the SOS Alliance] was that they were never able to transition from being an opposition group to being a part of a governing coalition."[99] A municipal staff member echoes this sentiment stating, "It used to be that people would look to SOS for guidance. People aren't asking that question too much anymore. A little more sophisticated approach would have let them remain a player."[100] Such insight suggests that the all-or-nothing approach of the SOS Alliance was very effective in building momentum for environmental protection initially but this oppositional approach is not a long-term political strategy, a classic weakness of social movements.[101] Although a threat to a local community can be a tremendous opportunity to galvanize widespread support, maintaining that support over time requires a transformation into more constructive forms of politics, something that the SOS Alliance has arguably failed to do.

The erosion of the SOS Alliance as a central player in Austin politics was evident in the May 2006 elections, when the organization and other environmental groups successfully placed a city charter amendment on the election ballot. Proposition 2 called for limiting municipal infrastructure and financial incentives for development over the aquifer as well as making all grandfathering decisions subject to city council approval. The amendment was essentially a call for the community to restate its commitment to the aquifer and water quality, but was soundly rejected by voters with 69 percent opposed, serving as a decisive loss to the SOS Alliance.[102]

At the same time, environmental politics in Austin over the last decade has evolved to embrace a more collaborative approach between environmentalists and developers. An environmental activist notes, "We're working on a new model of environmental protection with relationships as the core. The SOS approach is outdated and ineffective."[103] There are increasingly indications that the Austin environmental community is replacing the confrontational and divisive approach that emerged in the 1970s with a nuanced and compromising politics that reflects the current rhetoric of sustainable development. In other words, the singular focus on protecting the springs has gradually been supplanted by other issues such as air quality, urban gentrification, affordable housing, and energy use. The environmental activist simply states that "the focus on Barton Springs is necessary but insufficient to create a sustainable society."[104] An example of a new organization that reflects these values is Liveable City, formed in the early 2000s to address social, environmental, and economic issues simultaneously; its executive committee includes several former SOS board members who became disenchanted with the confrontational and litigious approach to environmental protection. Other SOS board members created the Hill Country Conservancy, a coalition of environmental and business interests dedicated to conservation land development approaches.[105]

Beyond the Austin City Limits

One of the most significant challenges of the SOS Ordinance is that it is only applicable within the City of Austin's municipal jurisdiction. An environmental activist states, "I think with the Barton Springs issues, there's really a sense that the questions have moved outside of Austin. Within our ETJ [extra-territorial jurisdiction], most of our development decisions have already been made, the die has been cast."[106] Thus, there are fears of leapfrog developments cropping up in neighboring jurisdictions that will affect water quality and result in a more sprawling metropolitan area. Similar to the clash between property ownership and environmental regulations mentioned earlier, there is a fundamental mismatch between the hydrologic flows of the Edwards Aquifer and the fragmented jurisdictional authorities that regulate growth over the aquifer.

To augment the regulatory approach, the City of Austin embarked on an ambitious conservation land development program. In 1998, voters approved $65 million in municipal bonds to purchase fifteen thousand acres of undeveloped land for water quality protection in the Barton Springs Zone. Additional bond money was approved by voters in subsequent years, and as of 2007, the City of Austin's Water Quality Protection Lands Program manages about twenty thousand acres in Travis and Hays Counties, 60 percent of which is under conservation easements and the remainder through full property ownership.[107] Effectively, these lands serve as enormous stormwater BMPs to protect water quality.[108] The conservation land development approach to protect water quality of the Edwards Aquifer differs from the regulatory approach by using market-based mechanisms that follow the 1980s mantra of the Nature Conservancy: "We protect land the old-fashioned way: we buy it."[109] An environmental activist notes that conservation development acknowledges that "property rights proponents have a right and the best way to fight it is to play the property ownership game."[110]

In effect, these jurisdictions have adopted the economic rationalist framework of the development community but reinterpreted it for environmental benefits rather than profit motive. Property can be purchased for the public trust as opposed to being a private commodity. An unintended consequence of purchasing land for conservation is that it inflates property values in the region and makes further conservation purchases more expensive. Environmental law scholar William Shutkin summarizes the problem with conservation land development as follows:

Simply purchasing open space with taxpayer dollars is the easy way out. It is much more difficult to undertake comprehensive planning that addresses both the built environment and undeveloped land. To the extent that states and municipalities buy up islands of land while leaving unaddressed the more complicated problem of how to build and develop in an ecologically sound manner, they will have achieved only a small success. Dollars for land acquisition are one thing; comprehensive, long-term planning for sustainability is another.[111]

Meanwhile, the impact of the water quality measures by the City of Austin (via regulatory and market-based approaches) and the federal government (via the Endangered Species Act) on the Barton Springs segment is difficult to quantify. Only a handful of properties have been developed under the SOS Ordinance; the majority of the land either has been developed under grandfathered regulations or remains undeveloped. An unpublished 2006 land use study by the City of Austin's Watershed Protection and Review Department shows that 46 percent of the Barton Springs Zone within the City of Austin's municipal jurisdiction has been developed, 31 percent has been designated as conservation land, and the remaining 23 percent has the potential for development (only a small portion of which can follow older, less restrictive water quality regulations). See figure 3.7.

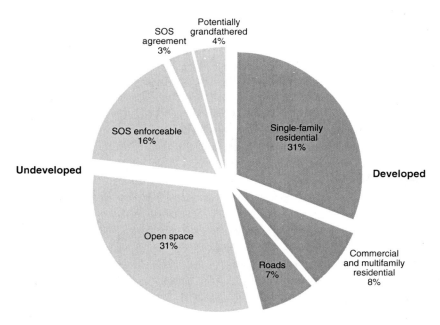

Figure 3.7
Land use in the Barton Springs Zone. *Source*: City of Austin 2006.

The study indicates that the fate of land development in the Barton Springs Zone within the City of Austin's jurisdiction is largely decided. Still up for debate is the remaining 72 percent of the Barton Springs Zone outside of the municipality where water quality regulations are less stringent or nonexistent. A stakeholder process for creating a regional water quality plan in the Barton Springs Zone was initiated in 2002 by a consortium of politicians and public officials in the region.[112] The elements of the plan are slowly being adopted by neighboring jurisdictions, notably Hays County, and the City of Dripping Springs, but it remains to be seen how Austin's water quality regulations will filter upstream through the Hill Country.

The Future of the Springs

The regulatory approach employed upstream of Barton Springs can, on one hand, be seen as a triumph by environmental protection advocates to reduce the impacts of urban expansion in an environmentally sensitive area. Austinites translated their cultural appreciation of Barton Springs into a stringent water quality ordinance that has created an uneasy but increasingly stable truce between green romantics and economic rationalists, at least within the City of Austin's regulatory jurisdiction. Outside of the municipality's jurisdiction, Austinites have funded a conservation land development program to extend their environmental protection goals. It is unclear if a large-scale, multijurisdictional approach will develop from the spillover effects of Austin environmental politics and dictate a form of regional development that can protect water quality.

However, both the regulatory and market-based approaches frame nature as outside of the human world. The fragile Hill Country landscape is interpreted in preservationist terms as a place to protect from human influence and preserve in its original state. This is reflected in the SOS Alliance's "either/or" proposition for the future of the Hill Country: pave it or save it. In such instances, hybridity and heterogeneity are kept at bay through robust topographic interpretations of the landscape that include property ownership and regulatory restrictions.[113] Notions of urban ecology are largely nonexistent in Austin despite the ability of water to transgress the market and regulatory boundaries that have emerged over the last three decades. Perhaps the spatial separation is a necessary stance given the sensitive geologic conditions of the Hill Country and the lack of certainty that land use restrictions and design strategies can offer to protect the region's ecological resources. In other words, it may be that we simply do not know how to live in such a fragile landscape without soiling the waters for downstream inhabitants.

Furthermore, the impacts on the environment are interpreted only in terms of urbanization and do not focus on other land uses, particularly agricultural practices that also contribute to erosion and water quality degradation. The purified landscape is one that is decidedly nonurban, but this fails to acknowledge the devastating impacts of rural landowner activities, specifically agricultural practices. The interpretation of all forms of urban development as "unnatural" has resulted in few examples of design intervention by municipal engineers and landscape architects to mitigate water quality impacts of new development. Emerging notions of Low Impact Development and source control strategies are not considered to be valid approaches to protect water quality in the region because all forms of urbanization are interpreted as harmful. Reflecting on the lack of innovation in urban development, a city hall insider notes, "The environmental community doesn't think so much about BMPs as a way to protect water quality. But in their defense, they don't know what sensible development in this town would look like. They don't have a positive model of land development."[114]

Two recent commercial developments in the Barton Springs Zone, Escarpment Village and the AMD corporate campus, utilize a number of source control strategies (green roofs, pervious paving, biofiltration strips, and rainwater harvesting) to meet the SOS Ordinance requirements, but these projects are not regarded by most as examples of best practices in water quality protection. Rather, green romantics interpret the source control strategies as superficial window dressing on unwanted new development projects that will ultimately serve as magnets for suburban development. Reflecting on the attitude of land developers and builders, an environmental activist notes that "they seem to think a green building is automatically beneficial, but this ignores the ramifications of having a big employment center over the watershed."[115] In other words, beneficial water quality strategies at the site level cannot address the larger aggregate impacts of growth in the region. Again, this suggests that individual properties cannot be considered in isolation but instead need to be understood in relation to the larger region in which they are located.

Conclusions

The population of Austin continues to expand while the prominence of Barton Springs as the defining locale of the community slowly diminishes. The City of Austin's demographer estimates that the municipal population will double in the next three decades, and an environmental activist worries about the relevance of the springs in the future: "With the huge influx of newcomers, the general awareness is diluted. It's very difficult to keep pace on educational outreach when the population is growing as fast as it is."[116] As such, it seems increasingly unlikely that the community will galvanize their collective energies around Barton Springs as they did on June 7, 1990. Reflecting on changes since the 1970s, another environmental activist notes:

I think the rapid rate of growth clearly dilutes the ability for a community to remain cohesive and maintain its understanding of its history and identity. When people come here from Dallas or Houston, they think it's great compared to where they came from. But they don't understand why we're so upset about some little development. So there isn't that same appreciation for what made it great and preserving that. Austin isn't beautiful compared to Houston and Dallas by accident, it's because everyone got out there and fought for it.[117]

In 2007, the Barton Springs pool was closed for a record number of days due to rainy conditions that caused flooding and the introduction of unsafe concentrations of contaminants into the pool.[118] Most agree that the water quality of the pool will continue to decline as new upstream development occurs; a respected local geologist recently made the controversial prediction that Barton Springs will be unswimmable in two decades.[119] This suggests that degradation of the springs—and, by extension, the unique culture of Austin—is an inevitable consequence of urban growth. As sociologist Scott Swearingen writes, "Barton Creek and Barton Springs are not dead yet, but as Austin gets even bigger and more growth covers the aquifer, they are more threatened than ever."[120]

The debate over environmental quality upstream of Barton Springs ultimately revolves around a conventional conception of nature as separate from humans. The solution to environmental degradation from this perspective is to stop urban expansion in the Hill Country, which is reflected in the all-or-nothing rhetoric of environmental groups such as the SOS Alliance. A larger dialog on what types of development might protect water quality in the region as a whole is nonexistent because of a singular focus on Barton Springs and the creeks and subsurface water flows that feed into it. Nowhere is there a topologic understanding of landscape that recognizes the interconnectedness of all activities in the region. In the next chapter, I head downstream from Barton Springs to see how the environmental politics of the Hill Country have spilled over into the central part of the city. Urban runoff strategies in the central core reflect a markedly different attitude toward human/nonhuman relations but are also implicated in the water quality activities described in this chapter.

4

After the Flood: Retrofitting Austin's Urban Core to Accommodate Growth

On Memorial Day 1981, Austin was subjected to a dramatic Texas-sized storm that produced ten inches of rain in a period of six hours. The hundred-year storm resulted in widespread flooding throughout the city and was particularly damaging to Shoal Creek on the original western boundary. The flow rate of the creek increased by an order of five magnitudes, from ninety gallons per minute to more than six million gallons per minute, creating a torrent of water and debris that scoured the creek channel as it raced to the Colorado River. Thirteen people lost their lives and property damage was estimated to exceed $35 million. The flood continues to serve as a contemporary reminder of the threat that urban water poses to the residents of Austin.

Austin's reputation as a progressive leader in environmental protection is largely founded on the municipality's activities to protect undeveloped land upstream of Barton Springs through strict water quality ordinances and conservation land development practices, as described in detail in the previous chapter. However, the municipality also has an international reputation as a leader in urban stormwater management that includes comprehensive monitoring activities, the early adoption of BMPs for new urban development, and the retrofit of previously developed areas of the city with state-of-the-art water quality controls. In this chapter, I head downstream from Barton Springs to examine the stormwater management practices of the municipality in the inner core of the city. Here, the challenges of urban runoff are markedly different from those of the Hill Country and involve a difficult balance of protecting residents and property from flooding and erosion while meeting water quality goals in a highly impervious landscape. The inner city stormwater activities in Austin provide a corollary to the political and administrative activities at the urban fringe, shifting the debate from nature preservation and future land development to retrofitting a built environment that is embedded with competing economic, cultural, and political meanings.

Metropolitan Nature in Austin

Since its humble beginnings in 1839, Austin has frequently been recognized as a city with great potential to develop in harmony with nature. A 1970s citizen's group notes that "the idea of a greenbelt city has captured the imagination of Austinites from the city's founding."[1] The first comprehensive plan for the city, the *1928 City Plan*, presented detailed plans for municipal parks along Austin's waterways because "the natural beauty of its topography and the unusual climate [make] it an ideal residential city."[2] In 1961, a newly adopted master plan introduced the concept of greenbelts or linear parks along the major inner city creeks and subsequent master plans continued to emphasize the importance of urban nature.[3]

One of the most fully realized visions of Austin's metropolitan nature is the bicentennial project of 1976, which proposed a comprehensive plan for the city with the creeks as organizing elements for recreation and transportation infrastructure. Both Shoal Creek and Waller Creek underwent civic improvement projects in the 1970s and 1980s to realize the greenbelt visions of various planning endeavors while providing flood and erosion control. However, the 1976 plan continues to be an idealized dream of integrating the creeks in the urban fabric. Construction and maintenance of creekside parks and hike-and-bike trails is intermittent and public access tends to be limited to short stretches of creek. Meanwhile, construction continues on the Shoal Creek hike-and-bike trail to undo the damage of the Memorial Day flood almost three decades ago.

Related to the local desire for metropolitan nature is a call to protect water quality. The issue of water quality has been a central aim of urban planning since the beginnings of the *Austin Tomorrow Comprehensive Plan* process in the mid-1970s, due in large part to the cultural importance of Barton Springs. A 1987 poll of the general population found that water quality trailed only traffic congestion as the most pressing problem faced by Austin residents.[4] The emphasis on water quality in the urban core is a product of the historical importance of the creeks and rivers to early residents, evolving knowledge about the state of the environment through scientific study of the region's water resources, and environmental degradation concerns that emerged in the 1970s.

The conditions of Austin's creeks today are strikingly different from those that attracted early residents. Shoal Creek and Waller Creek were once described as "teeming with fish" and having "reaches of deep water . . . [affording] choice bathing and swimming places for the population."[5] After the mid-nineteenth century, the creeks were subjected to the same utilitarian uses as other urban waterways in the United States. Rather than serving as the central organizing framework for the city, the waterways have been used primarily as conduits for wastewater and as repositories for urban detritus, notably transients and trash. In this sense, the creeks of Austin have become the natural counterpart to the city's alleyways, home to many of the undesirable elements of the city.

An engineer described Shoal Creek and Waller Creek in the early twentieth century as open sewers for both sanitary and stormwater flows.[6] With the introduction of comprehensive pipe networks to manage sanitary wastes in the 1910s and 1920s, the creek beds became the obvious location for sanitary sewer lines because they were at the lowest relative elevation (figure 4.1).[7] Furthermore, the stormwater networks were designed to discharge into the creeks.[8] Urban runoff from the increasingly impervious landscape was first collected in the drainage network and then deposited in the creeks and eventually the Colorado River. An Austin

Figure 4.1
An uncovered sanitary sewer in the bed of Boggy Creek in East Austin.

historian describes the storm drains sticking out of the banks of Waller Creek as mouths of cannons, poised to erase the prehuman attributes of the creek.[9] Today, the sanitary sewer and stormwater networks are an integral part of the urban creeks. They create a tightly coupled hybrid system of nature and technology, with stormwater flows mixing with the creek water and a subsurface pipe carrying sanitary waste volumes to the wastewater treatment plant downstream and east of the city.

A Hard Rain's a-Gonna Fall

The historic record of Austin reveals a long history of the municipal government and urban residents languishing due to the high frequency of devastating floods. The threat of flooding in Austin's urban core has resulted in a municipal strategy with a primary goal of protecting lives and property from periodic storm events. The flashy or volatile character of hydrologic flows is readily apparent in eroded or heavily armored creek banks, large concrete cobble, and trash lodged high up in the trees after severe storms. Walking along the creeks of Austin reveals the collusion between the impervious surfaces of the city and the violent hydrologic regime that has produced numerous floods in Austin over the years (see figure 4.2).[10] A staff member of the City of Austin's Watershed Protection

Figure 4.2
The 1915 flood of Waller Creek in downtown Austin. *Source*: PICA 04088, Austin History Center, Austin Public Library.

and Development Review department, the municipal department responsible for stormwater management, describes the conditions in Austin:

What you get when you have urban development is that the creek system just starts to unravel. What was once a naturally flowing creek gets shut off in our climate. The concrete and impervious cover doesn't allow for infiltration and contribution to groundwater flow and baseflow. So our creeks are drying up, they are eroding and the channels are widening because we have runoff at a higher velocity, we have more flooding because of the impervious cover and we have lower water quality. The forces of nature are having a huge effect.[11]

The flashiness of the hydrologic regime in Austin is compounded by an increasingly impervious landscape and a highly erosive soil horizon. The first street paving in Austin took place in 1905 on Congress Avenue, the grand promenade that connects the Colorado River to the State Capitol. Visitors are quick to recognize the large expanses of pavement in the city center, not only on Congress Avenue but throughout the downtown and adjacent neighborhoods, and particularly the transportation infrastructure that serves automobile and truck traffic. Today, the urban watersheds consist of about 50 percent impervious cover from roofs and paved surfaces, with some land usage (such as shopping malls and the central business district) almost completely impervious.[12]

The geology of the region exacerbates the highly impervious landscape of the inner core. The city is situated on a geologic transition zone called the Balcones Escarpment, which separates the Edwards Plateau to the west from the Texas Blackland Prairie to the east (see figure 3.1).[13] Unlike the Hill Country landscape, the Blackland Prairie to the east consists of gently rolling terrain with shallower stream valleys and deep soils. The large volumes of stormwater from impervious surfaces cut down the loose soils and widen the receiving waterways. For example, Little Walnut Creek in East Austin expanded from an average width of twenty feet wide to eighty-five feet wide between 1962 and 1997 (see figure 4.3).[14]

The Memorial Day flood of 1981 served as a reminder of the threat that urban waterflows pose to life and property. Responding to the flood damage, the city implemented a drainage fee; between 1981 and 1984, voters approved $75 million for capital improvement projects related to flood protection.[15] The result has been a preponderance of armored creek banks that prevent the creek channels from eroding, meandering, and causing property damage. Early, informal methods of bank stabilization involved dumping concrete onto the banks and letting it solidify. More recently, this practice has been refined to include formed concrete walls, gabion walls (cobble in wire cages), and geosynthetic fabrics that mimic vegetated creek banks. Though still a structural strategy for flood and

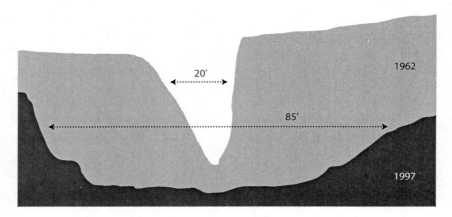

Figure 4.3
Cross-sectional diagram of Little Walnut Creek illustrating erosion over time. Illustration by Shawn Kavon, modified from City of Austin 2001b.

erosion control, vegetated walls offer a more pleasing aesthetic and provide at least a modest amount of riparian habitat. In the most successful instances, it is difficult to recognize that the bank has been stabilized.

The emphasis on flood control and armoring is frequently tied to ideas of economic development and urban improvement; a prosperous city is one in which the economy can grow on a stable, predictable foundation. This applied to the damming of the Colorado River in the 1930s and 1940s and continues on a smaller scale with the creeks in the urban core. A contemporary example of coupling water control with business development is the Waller Creek Tunnel Project, slated for completion in 2014. The creek was channelized in the 1970s to control flooding in the eastern part of downtown Austin and in the last decade, new designs for a stormwater diversion tunnel were proposed along with a master plan. Nearly a mile long and twenty-two to twenty-six feet in diameter, the $127 million tunnel will accommodate the majority of flows in the creek, even during extreme flooding events, to protect the surrounding buildings and streets.[16] A new pumping station will extract water from Lady Bird Lake and bring it to the top of the channel where it will be released to provide a constant surface flow reminiscent of a bubbling brook. The project will create an amenity for downtown businesses to attract tourists and shoppers, and is frequently compared to San Antonio's successful Riverwalk development.[17]

Flood control projects such as the Waller Creek tunnel demonstrate the continued embrace of the Promethean approach to controlling nature using the latest technological strategies. They create a sharp distinction between the built environment and predevelopment hydrologic flows, suggesting that the control of nature is the inevitable outcome of urbanization. However, such approaches to urban flood control often conflict with activities to improve water quality because stabilizing the channel for flood protection and erosion control is in direct opposition to restoration efforts and the creation of biological habitat. A Watershed Department staff member notes:

Internally our department is struggling with the mission of flood control and doing creek channelization without affecting water quality issues. I don't see how it can be done. The stream systems are so dynamic and in some places they're degrading or there's erosion happening and in other places, the land flattens out and it's dumping some of the bedload. So you really have to know where you are in the system and the changes over time in the drainage area due to urbanization.[18]

In hindsight, early municipal leaders would have done well to include significant setbacks from the creeks to keep structures and roads away from the eroding creek banks and the flashy water flows. A Watershed Department staff member notes that "in the inner city, some of these lot lines go right down to the centerline of these creeks. And the houses are right there. It's incredible what we are up against"[19] (see figure 4.4). In

Figure 4.4
A cantilevered building over Shoal Creek in the urban core.

limited instances, the municipality has engaged in the expensive practice of purchasing structures that are threatened by erosion and floodwaters, allowing urban waterways more room to meander within their channel.[20] These activities are the urban equivalent to conservation land purchases and easements in the rural Hill Country, pulling the urban fabric away from the unpredictable flows of nature. But in most cases, the municipality is stuck with a network that has evolved over the past century, with the built environment and nature entwined in configurations that tend to resist the large-scale changes required to improve water quality.[21]

Building a Municipal Stormwater Program

Despite the challenges of providing flood and erosion control in a flashy hydrologic region, the municipality has developed an internationally recognized approach to protecting and enhancing urban water quality. One of the most renowned activities is the longstanding monitoring of water quality conditions for watersheds throughout the city.[22] The earliest monitoring data in Austin was collected in the late nineteenth century, but the current program was initiated in the mid-1970s when the municipal government collaborated with the U.S. Geological Survey to monitor stormwater flows in eight large, mixed-use watersheds.[23] These monitoring activities preceded federal permitting requirements for stormwater management and were already well-established when Austin joined the federally funded National Urban Runoff Program in 1981 (as described in chapter 1).[24] A local geologist notes, "Austin already had ten years of data before this study came about, the city was way ahead of the curve. They had automated samplers on many urban creeks in the 1970s, which was unprecedented at the time."[25]

Although monitoring does nothing to alleviate the tensions between the built environment and water flows, it recognizes the value in tracking how environmental conditions change over time. In addition to monitoring activities, the municipality is well known for its early efforts to evaluate and require end-of-pipe BMPs for urban runoff from new developments. Austin established a national reputation in the early 1980s when it instituted water quality regulations that required high levels of treatment attainable only with sand filters (sometimes referred to as "Austin" filters). The regulations were performance-based and required specific pollutant removal levels. Sand filters were specified as the most cost-effective way to meet these requirements and eventually became the de facto standard for new development. The advantage of sand filters is that they provide a high level of treatment in a small footprint, as well as modest long-term maintenance requirements.

The early codification of high-efficiency treatment strategies is simultaneously a strength and a weakness of the municipality's water quality program. Although it represents a strong commitment to protecting waterways from urban runoff impacts, it also tends to limit the choice of treatment strategies. A consulting engineer notes, "The downside is that the high standards discourage innovation to some extent because you have to get variances to do anything that isn't approved in the manual, and variances are hard to get. So you don't see much experimentation here because the rules call for sand filter or equivalent."[26] In other words, the reliance on a proven technology as specified in the municipal regulations comes at the expense of innovation in treatment strategies. The regulatory structure creates particular expectations for the existing drainage system and a "technological momentum" to continue stormwater management in the established manner.[27] As a result, more recent source control strategies have not been adopted because of the difficulties in proving equivalence with existing end-of-pipe strategies.[28] In the mid 2000s, municipal staff members updated the city's *Environmental Criteria Manual* to include a handful of source control measures, including vegetated filter strips, rain gardens, rainwater harvesting, porous pavement, and biofiltration, but the majority of new development projects continue to rely on conventional strategies with established performance records. A Watershed Department staff member notes: "The development community wants to use [source control] techniques more than we do. I think if we already had green roofs, rainwater harvesting, bioretention, and porous pavement in our regulations, people would be using these techniques right now. They want to use it and they want to get credit for it but it's not very explicit. You could do something alternative but it's a hassle and it slows your project down, so no one wants to do it."[29]

The municipal water quality regulations stabilize the network of nature and society in the urban environment via well-established technological strategies, just as the watershed ordinances stabilize land development practices in the Hill Country. The codes can be understood as a means of translating the water quality interests of Austin residents into material practices, but the resulting system of regulations is obdurate, resisting modification and circumvention by newer strategies based on the assumption that they cannot meet existing pollutant removal thresholds. Although the establishment of the water quality program in the 1980s is interpreted by many as a significant achievement, evolution of the program in the intervening years to reflect new approaches to stormwater management has largely not occurred.

Furthermore, the municipality suffers from two significant deficiencies that plague all stormwater programs: inspection and maintenance.[30] It is one thing to monitor conditions over time and require treatment levels and quite another to ensure that existing systems are working as designed. The Watershed Department is charged with inspecting all BMPs annually and is responsible for maintaining systems for single-family residential development; commercial and multifamily property owners maintain their own BMPs. A Watershed Department staff member describes the problem:

Stormwater utilities across the country are faced with the issue of every time a new stormwater treatment facility is built, it has to be maintained. Here in Austin, the municipality takes responsibility for ponds and treatment systems that are residential. If it's for a private commercial development like an office building, the developer or owner is charged with maintenance. And multi-family is included in that too. We'll inspect these and we require them to be maintained. But we do the maintenance for single-family residential subdivisions and there are more of them coming on line every year. We have thousands to maintain. We don't want to increase our budget every year but we kind of have to keep up with that kind of stuff.[31]

The maintenance and inspection issues reflect the municipality's challenges in managing a complex and continually expanding technological network. The current drainage network of Austin comprises two hundred square miles and four thousand miles of drainage pipes, as well as thousands of flood and water quality controls.[32] Municipal staff is charged with managing the network using the latest scientific expertise within a bureaucratic hierarchy. A Watershed Department staff member sums up Austin's water quality program: "Everybody wants the image of Austin to be progressive and innovative and certainly not lagging behind but the reality is we're more reactive than our reputation suggests. When I go to Portland or Seattle or the mid-Atlantic states, they're doing a lot more cool stuff than we are. They're in a different place."[33]

The population growth issues of Austin described in the previous chapter strain the administrative capacity of the municipality to fulfill and maintain the ambitious goals of greening up the Promethean infrastructure of the past while maintaining the existing system. Ultimate control of nature requires energy and finances that are beyond most, if not all, contemporary government bureaucracies. Beyond the lack of staff and budget available for inspecting and maintaining these facilities, a more insidious problem is that the municipality does not even know the location of more than half of these structures. Reflecting on the challenges of maintaining such a vast system of drainage strategies, a municipal staff member notes, "We don't know where they are, we don't know how they're working, we have doubts about whether they are working well or not, and there's an inertia through the whole thing such that we continue to go forward."[34] To remedy the situation, in 2005 the department began a comprehensive field survey to identify all BMPs in the municipality's jurisdiction and construct a database to manage them efficiently and effectively. As of November 2007, the database contained about 4,700 commercial BMPs and 860 residential ponds and was about 85 percent complete.[35]

One of the most striking differences between the stormwater activities in the Hill Country and the inner city is the central role of municipal experts. Where environmental activists and landowners are the central actors upstream of Barton Springs, inner-city activities are dominated by Watershed Department staff members that rely on rational environmental management strategies that are scientifically proven and cost-effective. This difference reflects a bias of environmental activists who see the inner city as contaminated and unworthy of environmental protection as well as the heightened threats of floodwaters to private property and human life.

Balancing Flood and Erosion Control with Water Quality Goals

Austin's approach to managing urban water flows has changed over time, and with respect to water quality, the greatest transition has occurred in the last two decades, long after the municipality earned its reputation as a leader in stormwater management. In 1996, the municipal government underwent one of many departmental reorganizations and brought flood control and erosion control (originally in the Public Works Department) together with water quality protection (originally in the Environmental Resource Management group in the Planning Department). The newly formed department, Watershed Protection (later named the Watershed Protection and Development Review Department), has adopted an integrated environmental management approach that recognizes the interrelation of the various drainage mandates of the municipality. The departmental reorganization resulted in a substantial change in the administrative culture of the Watershed Department staff and emphasized a collaborative approach to reconcile the tensions between the various drainage missions. A Watershed Department staff member notes: "In 1996, it was basically two cultures coming together. The engineers from the Public Works Department were mainly focused on flood control and they got combined with the Environmental Management group where the water quality folks were. And there was this mix of cultures that really was a bit of a clash at first. In the last five years, we've tried to marry the two cultures and work together to produce integrated projects as opposed to single mission projects."[36]

Another Watershed staff member has a similar perspective on the new integrated approach of the department:

It was a gradual change but one of the main reasons for it was because many of the fixes for flood problems became environmental problems. To fix a flooded area, one of the obvious strategies is channelization or other approaches that ruin the natural system. And we butted heads on those things for years and years. If you work elbow-to-elbow with people, you come to understand why we think the way we do and why they think the way they do. Creating an environment where we could work together was a huge step for us and it took a long time to get there.[37]

The integrated approach is an attempt to find common ground between different missions within the municipal bureaucracy. Today, there is an explicit recognition that these goals can sometimes work at cross-purposes, and the new approach of the integrated department is to find strategies that can satisfy multiple aims simultaneously. Thus, the administrative rationalist model of environmental management in the Watershed Department has been transformed into a multidisciplinary collaboration that requires cross-pollination to find optimal scientific, technical, and economic solutions. It is a nascent recognition that environmental flows cannot be effectively managed in a segmented fashion because this merely pushes the problem from one geographic region to another or solves a flooding problem while creating a water quality problem. This is a common critique of rational environmental management approaches and will be addressed in detail in chapter 7.

One of the first major projects of the newly formed department was the ambitious *Watershed Master Plan* published in 2001.[38] The plan summarizes the flood, erosion, and water quality conditions in seventeen watersheds in the municipality's jurisdiction that represent 32 percent of the municipality's land area (including its five-mile extraterritorial jurisdictions) and the majority of the city's population. The study includes detailed analyses of the watersheds and overlays these analyses to produce an integrated prioritization of service needs. The authors provide a sobering cost estimate of $800 million over the next four decades to fulfill the Watershed Department's integrated goals, equivalent to about twice the historical capital spending rate.[39] The implicit message of the plan is that the municipality will need to make a substantial financial commitment if the simultaneous goals of flood control, erosion control, and water quality are to be achieved.

In the plan, the watersheds are characterized in three groups: urbanized, developing, and rural. The eleven urbanized watersheds are those with greater than 50 percent impervious cover and most of the creeks in these watersheds have more than doubled in size due to increases in urban runoff. Future enlargement of the channels in these watersheds is expected to be low, indicating that these systems have already expanded and are relatively stable. Integrated strategies in these watersheds include channel restoration projects to reduce erosion and improve water quality, property buyouts, and upgrade of existing flood control and water quality ponds.[40] Conversely, it is the rural and developing watersheds where the majority of change will occur, as impervious surface coverage and scouring waterflows are expected to increase dramatically.

The plan offers mixed prognoses about each of the urban watersheds, although the lower reaches are generally in need of immediate intervention, due to more intensive land use and the cumulative impacts of urban runoff volumes from the watershed headwaters. As a whole, the plan calls for a piecemeal approach to target areas that are most problematic (and often most costly) to be addressed first. There are no comprehensive planning visions similar to the greenbelt proposals of the past, and the application of small-scale source control strategies is mentioned only in passing. There is a significant emphasis on the evolution of the existing drainage network to alleviate the issues of a network under continual crisis. One of the most interesting aspects of the plan is its explicit assessment of achieving its water quality goals in the inner core. Among the three missions, the Watershed Department estimates that water quality is the least likely to be achieved. The authors write, "Attainment of erosion and flood goals may be possible with sufficient funding. Water quality goals are not attainable through implementation of solutions evaluated in the Master Plan. Limited regional retrofit opportunities in urban watersheds and inadequate regulatory controls in areas outside of the City's jurisdiction are significant constraints."[41] In other words, lack of space in the urban landscape for land-intensive water quality controls in conjunction with no control over upstream runoff volumes prevents the Watershed Department from achieving its water quality goals in the urban core. Nowhere is there a romantic environmentalist notion of returning the waterways to predevelopment conditions; instead, there is an acknowledgment that in many cases, channelization is the best and only option.[42] Like the upstream dialog on land use, there is something of a fatalist attitude to water flows and the inevitable decline of the environment over time. A Watershed Department staff member notes, "Part of it is throwing up your hands and realizing that it ain't gonna look like a natural system. Sometimes you just have to go to a highly armored solution because it's just too much impervious surface to handle the flows in these creeks. So in some cases it's a giving up and realizing that we're not going back to nature."[43]

The Limits to Improving Urban Water Quality

One might assume that given the municipality's international reputation among water quality professionals and its newly integrated department for stormwater management, there would be a plethora of groundbreaking projects to visit throughout the city. However, beyond the scattered network of sand filters and the subtle but significant changes in bank stabilization approaches as noted earlier, examples of the municipality's commitment to water quality are few and far between. One of the most visible examples of the Watershed Department's integrated mission and most effective means of retrofitting the central core is its wet pond program.

A wet pond sounds like a redundant label for a naturally occurring waterbody, but in this context, it is a highly engineered structure that looks and acts like a natural water filter. Unlike detention basins, wet ponds have a permanent pool of water as well as wetland plants that remove nutrients and dissolved contaminants from urban runoff. The ponds serve as flood control devices, water quality filters, and aesthetic amenities to neighboring residents. The most widely known example in Austin is the highly successful Central Park wet pond in the Waller Creek watershed in the central core (see figure 4.5). This public/private partnership between the City of Austin, the State of Texas, and a private developer consists of a ten-acre park and a series of constructed ponds to manage runoff from 172 acres of landscape with typical impervious cover of 54 percent.[44] One Watershed Department staff member states:

Our group is pretty high on wet ponds because they are an aesthetic benefit, there is a reasonable chance at getting some nutrient control, as opposed to sand filters that don't do a good job with dissolved constituents. It's the use of natural systems but it's still end-of-pipe. In a way, it's kind of like bioretention or natural systems because you're taking advantage of nature's ability to process all of these pollutants. You're not just saying, "Let's create a miniature wastewater treatment plant with a sand filter."[45]

Figure 4.5
A portion of the Central Park wet pond in the Waller Creek watershed.

Wet ponds apply the logic of nature to a technical strategy for water quality control and reflect the nascent discipline of ecological engineering (as mentioned in chapter 1). The limitation of wet ponds is their requirement for large land areas and high costs for construction and maintenance. Design and construction of the Central Park wet pond totaled $584,000 and annual maintenance costs the municipality $35,000, although this reflects a special contract with the property owner. Several dozen wet ponds are now online in the municipality's jurisdiction but siting additional ponds is increasingly difficult, due to the large space requirements. A Watershed Department staff member notes, "It's a real challenge to find a site where we can shoehorn in a control and have a big enough drainage area to have it make sense from a cost-benefit perspective."[46]

One of the most intriguing aspects of the wet pond approach is its ability to blur the boundaries between technical, natural, and social systems. Rather than being a fenced-off or underground structure, the wet pond is a highly visible and aesthetically pleasing stormwater treatment facility. It rejects the conventional logic of separating urban residents from stormwater flows; although it continues to be a technomanagerial approach to urban runoff, it begins to recognize urban water flows as an opportunity for place making. However, this idea of place making and expanding stormwater management facilities as an asset to urban residents cannot be achieved by the Watershed Department alone. Integrating the drainage missions in the Watershed Department has been a significant evolution of the municipal approach to urban drainage, but extending this approach to include other departments has been less successful. Collaboration between municipal departments that are directly engaged with or affected by urban water flows—including the Watershed, Parks and Recreation, Transportation, and Neighborhood Planning Departments—is a rare occurrence.[47] A Watershed Department staff member notes:

We are finding more and more often that the Watershed Protection and Review Department wants to collaborate with Parks and Recreation, because so often municipally owned open space is in parks or greenbelt. We have to ask permission to use that open space for a water quality treatment system and if the runoff is going through active recreation areas like ballparks and things like that, we're generally not going to get permission because parks doesn't want to lose that function.[48]

The "mission focus" of the municipal departments creates a series of bureaucratic silos that are easily transgressed by water flows but not by municipal staff members who are interested in finding synergies between departmental missions. Restrictive institutional and disciplinary norms and structures continue to favor a topographic interpretation of how nature infiltrates the urban landscape.[49]

Growing In Rather than Out

Amid the struggles in the Watershed Department to balance flood and erosion control with water quality in the urbanized inner core of the city, the growth politics of the rural Hill Country have influenced these practices. Efforts by environmental and community activists as well as the City of Austin to limit development impacts in the Hill Country through regulation and conservation development have pushed urban growth not only outside of the municipality's jurisdiction but also to less environmentally sensitive parts of the city, particularly the eastern neighborhoods downstream of Barton Springs. As described in the previous chapter, the *Austin Tomorrow Comprehensive Plan* of 1980 prescribed a strategy to direct growth to the inner core of the city by making infrastructure upgrades, but fiscally conservative voters rejected subsequent bond measures intended to fund these upgrades.[50] A second comprehensive planning effort, the *AustinPlan*, was completed in the late 1980s in an attempt to provide the *Austin Tomorrow Comprehensive Plan* with regulatory teeth, but the development-friendly city council refused to ratify it. There was and continues to be an allergy to comprehensive city planning in Austin, due to the long historical tradition of property rights and a state government that tends to favor market-based approaches to urban growth.[51] Official municipal policy since 1980 has been to direct growth away from the Hill Country, but this policy has not been supported by a comprehensive land development plan due to financial, regulatory, and technical barriers.

Amid the local and state political battles over the legality of the SOS Ordinance in the mid- to late 1990s, the City of Austin instituted an incentive-based geographic solution to steer growth away from the Hill Country using principles of Smart Growth. Smart Growth is a multifaceted planning strategy adopted by numerous municipal, county, and state governments in the United States to strengthen the municipal tax base, reduce the cost of government services, create economically and socially integrated neighborhoods, provide affordable housing, protect farmland and habitat, improve local air and water quality, support local businesses, and increase neighborhood safety.[52] Smart Growth strategies are realized through incentives, market strategies, and land use regulations that are framed as positive rather than restrictive to urban development; urban growth is not questioned but is rather directed in particular formations, largely through land use policies and physical design strategies. Indeed, framing growth in "smart" terms is a rhetorical strategy to garner favorable support for planning and land use decisions that are typically interpreted as governmental restrictions.[53]

Austin's Smart Growth Initiative (SGI) is unique because one of its principal goals is to protect water quality (see table 4.1). SGI incentives include reduced development fees, utility reimbursements, and code amendments

Table 4.1
City of Austin Smart Growth Initiative goals

1. Determine how and where to grow	Steer growth to the center of the city with Traditional Neighborhood Development, Transit-Oriented Development, and other mechanisms
2. Improve our quality of life	Preserve and enhance existing neighborhoods, protect environmental quality, improve accessibility and mobility, and strengthen the economy
3. Enhance our tax base	Invest strategically, use public funds efficiently, and form regional partnerships

Source: City of Austin 2007b.

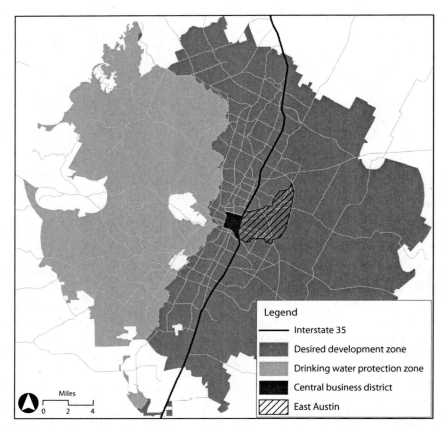

Figure 4.6
Map of Smart Growth Zones with the Central Business District and East Austin at the Core of the Desired Development Zone.

that encourage development in particular parts of the city. The centerpiece of the SGI is the Smart Growth map identifying a Drinking Water Protection Zone in the west and a Desired Development Zone in the east (figure 4.6). An environmental activist observed that "the designation of the Drinking Water Protection Zone and the Desired Development Zone simplifies the debate and clarifies it, for the development community and the citizens."[54] An important part of SGI was the framing of both Smart Growth zones in positive terms. It suggests that urban development can be a "win-win" proposition for growth machine advocates and green romantics by simply steering future growth to desirable locations.[55] The municipal program also reflects the urban sustainable development discourse that emerged in the 1980s and 1990s to simultaneously reduce sprawl, protect environmental quality, and develop a vibrant, mixed-use urban core.[56] The implicit message is that urbanization does not need to be a zero-sum game—it can be beneficial to both humans and nonhumans if done prudently and thoughtfully.

The Watershed Department's 2001 Master Plan explicitly supported the SGI strategy for combating sprawl and called for the "use of engineered strategies in the Desired Development Zone to minimize the need for additional regulations that may restrict development in that area."[57] As such, the SGI program serves to strengthen the Promethean approach to stormwater management in the urban core. However, the support of the SGI by the Watershed Department is most likely an act of political expediency and a show of support for a citywide development vision rather than a benefit to the Watershed Department's goals, particularly those related to water quality improvement. As noted previously, the plan states that the water quality goals will likely not be met in the inner city due to a lack of space to site, construct, and maintain BMPs, and that densifying the inner core will only make this task more difficult.

From a water quality perspective, the strategy of directing growth away from the Hill Country simply shifts the impacts of urban development from one locale to another while reifying the modernist dichotomy of city and wilderness. The SGI explicitly specified a zone of nature protection and a zone of nature destruction. A Watershed Department staff member reflecting on the implications of the SGI notes, "It means you are kind of giving up on water quality in the central part of town to protect some of the areas that are further away from the center parts of town. A lot of times, in an urbanized area there is no space to put in some kind of water quality treatment system. If you densify, you just can't put in those water quality controls in certain areas."[58]

Beyond the water quality and technology issues of densifying the urban core, the SGI has significant social ramifications, particularly for those residents whose neighborhoods are targeted as the Desired Development Zone. The SGI was embraced by green romantics and growth machine advocates in Austin but generated fierce criticism from low-income residents in East Austin, home to the majority of Austin's African American and Latino residents since the early twentieth century. The experiences in Austin with Smart Growth reflect larger concerns about these strategies that generally further market preferences without addressing the underlying causes of the problems they are trying to solve, namely sprawl, congestion, economic segregation, and environmental injustice.[59] The SGI's principal goal of protecting water quality pits environmental protection against social equity.[60]

East Austin's development can be attributed to both cultural and environmental influences. The city historically developed with minority populations to the east of Waller Creek and downriver from the rest of the city, where light industrial activities were intermingled with residents on smaller, more affordable lots.[61] The creek served as a natural delineation between Anglo populations to the west and African American populations to the east (Latino populations resided on both sides). As Austin grew, the segregation of the city was institutionalized through comprehensive planning documents and infrastructure provision. The most blatant instance of spatial segregation is summarized in the *1928 City Plan* prepared by the Dallas engineering firm Koch and Fowler, which noted the inefficiencies of providing dual municipal facilities for Anglo and African American populations throughout the city. The plan recommended that services for African Americans be restricted to the largely minority neighborhoods of East Austin.[62] Subsequent municipal policies promoted an increasingly segregated city while being careful to avoid unconstitutional municipal regulations. The western push of urban development beginning in the 1970s and the related water quality politics can be attributed in part as a desire by middle- and upper-income Anglo residents to be located away from East Austin.[63]

Similar to its population, East Austin's hydrologic conditions differ significantly from the rest of the city. Its location downriver from the central business district and at the bottom of several highly urbanized watersheds effectively makes East Austin the hydrologic drain for the city. Development of the creek headwaters in this part of the city began in the 1950s and peaked in the 1980s and 1990s, resulting in severe flooding and erosion problems downstream. These erosion problems were exacerbated by urban development patterns that situated buildings on small lots close to the creeks. In the late 1980s, the U.S. Army Corps of Engineers channelized East Austin's main waterway, Boggy Creek, to stabilize the creek slopes and protect residents and dwellings from harm (see figure 4.7). A local engineer states: "These channelization projects came about using the engineering of the day; there wasn't much thought about any kind of geomorphic constraints or habitat—it was just pure flood control to get the stormwater in a channel as small as possible and get it out. By having a smaller channel, you have less easement to buy, which affects project costs. The concrete trapezoidal channels were a normal way of business for flood control."[64]

Figure 4.7
The channelized portion of Boggy Creek in East Austin.

An environmental activist familiar with East Austin hydrology makes a more explicit connection between the channelized East Austin creeks and social and political conditions of the city:

In West Austin, we have big parks, big rights-of-way for creeks. In East Austin, we have some parks, those are areas that you know about. But if you start to walk Fort Branch or Little Walnut, you'll go by many creeks where the width of the right of way is maybe 50 to 100 feet. And so to get that flood flow down that right-of-way, you're not going to have a natural meandering channel, you're going to line it with a concrete trapezoidal channel and it's going to be dead. And that's just a matter of when the developer came in, how much right-of-way did the city make him dedicate to the drainage easement? And yes, that does reflect the price of the lot and the price of the house. But it doesn't matter which comes first: it just is a fact that the public resources are not as nice on the poorer side of our town.[65]

From an East Austin perspective, the SGI can be interpreted as another in a long line of municipal policies to dump the city's problems on East Austin residents. In this case, the problem was not a matter of undesirable minority residents or industrial activities but rather how to balance the municipality's competing interests of encouraging economic growth while satisfying the desires of environmental activists to steer this growth away from the Hill Country. East Austin became the most logical target for future population growth of the city.

Environmental Justice: An Alternative Environmental Discourse

In the early 1990s, environmental justice emerged as a potent political movement in East Austin. Echoing the Sanitary Movement of the nineteenth century, environmental justice advocates recognize the tight coupling of environmental conditions and human health.[66] In East Austin, a neighborhood group called PODER (People in Defense of Earth and her Resources) formed to fight the expansion of a gasoline tank farm and its potential health impacts on neighboring residents. PODER and other neighborhood organizations were joined by environmental groups including Clean Water Action, Earth First!, and the Sierra Club and successfully ousted several multinational petroleum companies from the tank farm property.[67] The Tank Farm Controversy was a significant victory for East Austin residents, due in part to a "red-green coalition" between environmental justice activists in East Austin and green romantics in Central Austin. The environmental justice activists returned the favor by supporting the passage of the SOS Ordinance in 1992, and the integrated coalition was seen as a "new rapprochement between minority and environmental groups."[68] The groups shared a common cause in the fight against the urban growth machine of developers and the municipal government.

However, the red-green coalition would last for only a few years; the introduction of the SGI and its promise to accelerate the gentrification processes already occurring in East Austin was an important element in the unraveling of the understanding of urban growth as both an economic and equity issue.[69] East Austin activists recognized the SGI as a way to diffuse the tensions between environmental and development interests in the Hill Country at the expense of low-income minority residents in the inner city. A PODER report does not mince words in characterizing the SGI as a municipal policy of disguised gentrification that allows environmental protection goals to propagate "internal colonialism" and a continuation of the urban renewal programs that affected East Austin in the 1960s.[70] A local newspaper columnist summarizes the unintended consequence of the SGI spatial fix: "Problem is that folks in the Desired Zones were never asked if they desired to be Desired. Smart Growth can be seen as a way to dump growth into neighborhoods that don't want it or can't handle it or are too disadvantaged or disempowered to do anything about it."[71] An environmental activist echoes this perspective, arguing that "the mistake in the 1990s is that we ignored social problems, we didn't relate the social problems of East Austin with environmental issues."[72]

Several respondents noted that property developers were instrumental in breaking up the red-green coalition after the passage of the SOS Ordinance, and one environmental activist goes so far as to claim that prominent developers "gave money to Eastside organizations and convinced them that Barton Springs was not an issue about them. This prevented the environmental community from creating a coalition with the social justice advocates of East Austin."[73] Although the role of development interests in breaking up the coalition between environmental justice and environmental protection advocates is a matter of debate, the protection of Barton Springs and the Hill Country was increasingly understood by environmental justice activists as an issue that would serve middle- and upper-income Anglo residents while having no appreciable benefit to low-income minority residents in East Austin. Saving the springs was an Anglo issue and its goals would be detrimental to the Latino and African American communities living in East Austin.

In 2003, the city council abandoned the SGI as official municipal policy due to a combination of issues, including a national economic downturn that slowed development activity as well as persistent and vocal protests by East Austin environmental justice advocates. There was a lack of enthusiasm from the development community for the market-based incentives offered by the municipality and there was unanticipated political fallout in foisting urban density on East Austin. Smart Growth was not a panacea for Austin's urban growth problems as initially advertised because it exacerbated the existing social tensions between different inner-city populations. Although the official program of inner core densification has been abandoned, the municipality continues to encourage more density in the urban core through a variety of less spatially specific strategies, including Transit Oriented Development, density bonuses that allow developers to exceed building height limits, and tax incentives to lure commercial businesses to the Desired Development Zone.

An unintended consequence of directing future growth to the city's central core has been a reassessment of the function and value of the inner-city waterways, specifically as they relate to infrastructure services. The vestiges of inequitable infrastructure provision between Central Austin and East Austin still exist, but several respondents noted that the attitude of the municipality has changed in the last two decades. A Watershed Department staff member notes: "We're very sensitive to eastside/westside issues. For instance, we look at whether we have as many ponds on the eastside as on the westside and the same with streambank stabilization. Are we doing the same type of projects and the same quantity of stormwater projects on both sides of the city? It's traditionally been a concern. And in the past, there may have been some divergent approaches but I haven't seen any of that since I started working for the City."[74]

An example of the new commitment to environmental quality in East Austin is the 2006 municipal purchase of a fifty-nine-acre lot that encompasses a natural spring. A city council member notes, "Our commitment to the environment is not just in one area of the city. Considering how much we'll spend west of I-35 and even outside of Travis County, this is a small request."[75] Reflecting on the project, a PODER representative states, "We're really pleased because we've always said we haven't gotten our fair share for protecting the environment and wetlands and green space in East Austin."[76] The rise of such constructive projects in East Austin suggests that the historic inequitable distribution of infrastructure service is slowly being erased.

Conclusions

Urban runoff activities in the inner core of Austin are strikingly different than in the Hill Country. There is no dominant citizen push to protect natural resources and no hope for achieving nondegradation of the heavily modified waterways in the inner city. Rather, the urban core dialog is about how much degradation is acceptable and how much the city's residents are willing to spend to make their creeks nonthreatening and at least somewhat healthy and aesthetically acceptable. This dialog is one of administrative rationality that focuses on directing municipal funds as efficiently as possible to solve flooding and erosion problems and—to a lesser extent—water quality. These activities are done on a piecemeal basis and tend to focus on areas in crisis, due to limited funds available for water quality activities and the enormity of urban water problems.

The imbroglio of nature and society is readily apparent in the inner city, as demonstrated by conflicts within the Watershed Department and among municipal departments, the condition of the creeks, and the political battles over the SGI in East Austin. As a whole, these activities reveal a lack of administrative capacity to resolve the tension between nature and humans. However, Watershed Department staff members are surprisingly optimistic and see their work as gradually improving the physical conditions of the inner core both socially and environmentally. A Watershed Department staff member states:

It may be 15 or 20 years before we see the fruits of our labor. And we may just be tamping down the rate of degradation and holding the line so things don't bottom out. One of the things that people don't have a sense for is the incredible level of effort that it takes to protect the inner-city creeks. Some people look at our regulations and see them as extravagant, but it is really hard to protect water quality. And just dealing with the historical repercussions of what we didn't used to know—it's mind-boggling, not only pollution control but volume control and rate of flood flows to protect streams. It's huge.[77]

The urban creeks continue to offer glimmers of hope for creating more desirable relations between humans and their material surroundings (see figure 4.8). Despite the eroded creek banks, the exposed sewer lines, the illicit dumping, and all manner of neglect and abuse, walking along the creeks—even in the lower reaches of highly urbanized watersheds—can at times be an enlightening and encouraging experience. A Watershed Department staff member notes:

It's amazing, walking these creeks you can see the ravaged conditions and then you turn a corner and you find an outcrop that has real beauty. There are places like that all over Austin, even in our most damaged creeks. And the thing that is encouraging for me is that people just love those creeks. They're bombed out, they're urbanized to the max, they are a humongous challenge to restore (and we're doing it slowly), but people love them and they'll fight to protect them. For me, that's very satisfying that people still see the beauty and still connect to those waterways. That makes our job worth it. It might be another 50 years before we

Figure 4.8
An inspirational spot on Blunn Creek in Central Austin.

fix Shoal or Blunn [Creek], but it's just a matter of getting the boat going in the right direction.[78]

In the flood-prone environment of Central Texas, it is perhaps most realistic to have modest goals for water quality, particularly in the inner city, where the built environment exacerbates the already flashy conditions. Despite the trashed conditions of the inner-city creeks, urban residents and municipal staff members continue to hope that the creeks can be assets rather than liabilities for Austin's future. This hope for an improved future drives municipal staff members to work toward shared missions of multifunctional and aesthetically pleasing stormwater structures and brings neighborhood residents back to their local waterways with desires to rework them into the fabric of their neighborhoods. In the meantime, the creeks in the inner city wait quietly for the next Texas-sized storm.

5

Metronatural™: Inventing and Taming Nature in Seattle

Metronatural™
adj. 1. Having the characteristics of a world-class metropolis within wild, beautiful natural surroundings. 2. A blending of clear skies and expansive water with a fast-paced city life.
n. 1. One who respects the environment and lives a balanced lifestyle of urban and natural experiences. 2. Seattle

In October 2006, the Seattle Convention and Visitors Bureau unveiled a new "brand platform" to market the city to tourists and convention goers. Metronatural™ serves as both an adjective to describe the perceived character of the city as well as a noun to embody the city and its residents.[1] The campaign was immediately panned by locals who thought it sounded like a contemporary nudist camp or was uncomfortably close to words such as "menstrual" and "metrosexual,"[2] but the concept reflects a widespread belief by residents and outsiders alike that Seattle is a city defined by, and in harmony with, nature. Indeed, regional writer Jonathan Raban argues that Seattle is the "first big city to which people fled in order to be closer to nature."[3]

In this chapter, I examine the evolving interpretation of nature in Seattle and the Pacific Northwest from the nineteenth century to the 1960s, a period when urban boosters invented an ideology of "metropolitan nature" to attract new settlement.[4] Meanwhile, Promethean activities by economic boosters and the fledgling municipal government would "improve" upon the promised landscape to facilitate economic and cultural development. The taming of big nature in Seattle produced some of the advertised benefits of social and economic progress but also had unintended consequences of exacerbating social inequality and creating a more unstable landscape. From this perspective, Metronatural is an apt description for the cognitive and material creation of a hybrid landscape in which humans and nature exist in an uneasy and continually shifting truce.

Seattle: Metropolis in the Promised Land

Seattle is a large metropolis of 560,000 residents located in the north-west corner of the continental United States (see figure 5.1). Despite the northerly latitude of 47 degrees, Seattleites enjoy a temperate climate due to the city's proximity to the Pacific Ocean.[5] Situated on a narrow isthmus between Puget Sound and Lake Washington, Seattle's rolling hills, abundance of water, and distant mountain peaks provide a stunningly picturesque locale for residents and visitors alike. Local historian Roger Sale writes, "Seattle is a soft city, made up of soft light

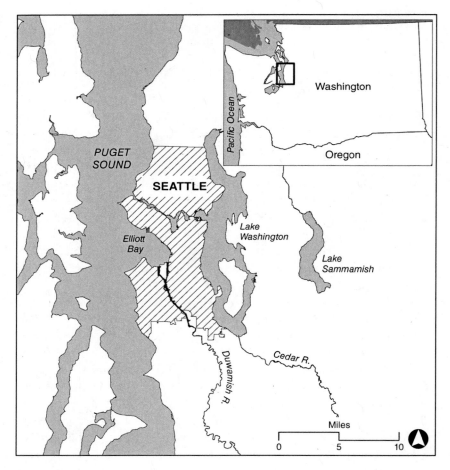

Figure 5.1
Location map of Seattle.

and hills and water and growing things and a sense that nature is truly accommodating."[6]

Seattle is defined first and foremost by its location in the heart of a region referred to as Cascadia, the Great Northwest, the Great Raincoast, God's Country, the Promised Land, Ecotopia, or, most frequently, the Pacific Northwest. The geographic extent of the region varies depending upon one's emphasis on climate, economics, culture, or a combination of the three. Historian Carlos Schwantes argues that "any search for commonly agreed upon boundaries for the Pacific Northwest will prove fruitless. . . . The regional perimeter, except along the Pacific Ocean, remains as indistinct as a fog-shrouded promontory."[7] All regional definitions include the coastal portions of Oregon and Washington as the western border, but the other boundaries are contested. Sometimes the northern edge extends to southern Alaska, and the southern extent can reach San Francisco and stretch as far east as Montana. Despite a lack of consensus on the region's geography, Schwantes notes: "In the Pacific Northwest, as in few other parts of the United States, regional identity is almost wholly linked to natural setting. The Pacific Northwest without its mountains, its rugged coastline, its Puget Sound fogs, its vast interior of sagebrush, rimrock, and big sky is as unthinkable as New England without a Puritan heritage, the South without the Lost Cause, the Midwest without its agricultural cornucopia, or California without its gold rush mentality."[8] And a specific element of nature—water—is at the heart of this regional identity. Historians Thomas Edwards and Carlos Schwantes write: "Beaver and evergreen symbolize the states of Oregon and Washington respectively. But an equally appropriate symbol for both might be a drop of water. In the Pacific Northwest as in few other regions of the United States, water is an abiding, if not always appreciated, presence; fog and rain, glacier and waterfall, irrigation canal and tidal estuary, the Columbia River and Pacific Ocean."[9]

Throughout the twentieth century, water has been understood as a significant driver of the regional economy, particularly during the storied infrastructure development period from 1930 to 1970 when a plethora of hydroelectric dams were constructed to harvest "white coal" as an inexpensive electricity supply.[10] The prominence of hydroelectric dams in the region suggests that the so-called Promised Land was an unfinished one requiring large engineering projects to perfect the region and make it habitable for human settlement. Furthermore, the dam projects—largely initiated and funded by the federal government—solidified the notion of the Pacific Northwest as a distinct region through the promotion of a rational and comprehensive planning logic. The inexpensive electricity from the dams was applied to the national defense industry during World War II and the Cold War, transforming the Pacific Northwest from a sleepy region into an international economic center.[11] The idea of the region was founded not only on its remote character and intrinsic beauty but also on the potential of the landscape to be exploited for economic gain.

In Seattle, water is a central component of its picturesque beauty. Puget Sound is the most prominent aqueous neighbor, forming the western boundary of the city that serves to "convey a sense that water is always near."[12] Urban planners Roberto Brambilla and Gianni Longo extend this idea in their description of the city: "Ideally situated on top of seven hills, 80 percent surrounded by water, with spectacular views of Lake Washington on its eastern flank and equally spectacular views of Mount Rainier, the Olympic Mountains and island-dotted Puget Sound, Seattle is a city with so much intrinsic beauty, so many breathtaking sights which can be seen just by looking out the window, that it starts with an advantage over most cities."[13] Equally important to Seattle's distinct culture is atmospheric water in the form of rain.[14] Situated between the Olympic Mountains to the west and the Cascade Mountains to the East, Seattle receives consistent rainfall about 150 days per year.[15] The resulting maritime climate is distinct from the arid inland plains to the east of the Cascade Mountains.

Early accounts of the Pacific Northwest by explorers and settlers, notably Lewis and Clark, included vivid descriptions of the unrepentant and depressing rainfall that plagued the region. The most memorable of these accounts described the coastal conditions to the west of the Olympic Mountains, where annual rainfall in excess of a hundred inches is common.[16] At the end of the nineteenth century, regional boosters debunked the notion of incessant rainfall by developing an "ideology of climate" to reframe the precipitation patterns as a blessing rather than a curse.[17] The campaign emphasized the advantages afforded by the climate of Oregon and Washington territory, particularly its mild and equable qualities. Though it was noted that a small portion of the coastal areas of the Pacific Northwest were subject to overabundant rainfall, the majority of the region enjoyed a climate similar to England, France, and Japan and could provide residents with an equally prosperous future.

Indeed, Seattle's annual rainfall of thirty-six inches is comparable to that of Dallas, Kansas City, Chicago, and Cleveland and is only half that of wetter cities such as Miami and New Orleans. And depending on how you define rain, one could argue that it hardly rains in Seattle at all; over two-thirds of rain events are less than 0.04 inches per hour and constitute nothing more than a hard drizzle.[18] What defines Seattle's precipitation pattern from other cities is its temporal distribution that results in an incessant light sprinkle, sometimes referred to paradoxically as a "dry rain."[19] A local newspaper columnist writes:

> It's not the amount of rain that defines the Northwest. It's the persistence. Our rain is a relative you thought you knew until the day he showed up on your doorstep. He came in for the night, stayed through the weekend. Monday, he missed his plane. By Thursday, he had migrated from the spare room to the kitchen to the living room, devouring space as he went. Pretty soon he'd taken control of the refrigerator, the television and the stereo. Eventually, it dawns on you, he's taken over your life. Rain moves in and sets up shop in the imagination.[20]

This mental preoccupation with rain by Seattle residents involves not only precipitation but the unavoidable cloudy and dark conditions. Seattle is in the Puget Sound Convergence Zone between the Cascade and Olympic Mountains, where clouds and rain are trapped for three-quarters of the year.[21] A journalist notes that "if the climate were not so dark and rainy . . . everyone would want to live here."[22] The bright side of Seattle's climate is that the moderate conditions also result in a lack of severe meteorological events such as droughts and flooding, although long-term risks of earthquakes and volcanoes are a continual threat.[23]

The preponderance of water in Seattle and the Pacific Northwest also touches on another important defining element of the region: salmon. The salmon is a totem of the region, similar to the crab of the Chesapeake Bay, the walleye of the Great Lakes, and the lobster of New England.[24] Writer Timothy Egan has controversially defined the Pacific Northwest as any place where salmon can reach.[25] Historian William Cronon adds that it is impossible to separate the species from the region: "The salmon is not just a keystone species but a cultural icon of the first order, a powerful symbol of all that the Pacific Northwest is and has been."[26]

With the introduction of canning technologies in the late nineteenth century, salmon became an economic opportunity of global proportions for Pacific Northwest fishermen; people throughout the world could now sample a taste of the region.[27] Industrial-scale salmon harvesting, along with the destruction of habitat for hydroelectric dams and urban development, resulted in significant declines in salmon populations; by the turn of the twentieth century, Pacific Northwest salmon populations were on life support, with wild populations sustained and supplemented by human-managed fish hatcheries. Historian Joseph Taylor refers to the challenges of maintaining salmon populations in the Pacific Northwest as an "enduring crisis" that continually attempts to reconcile human connection with and destruction of this iconic megafauna.[28] Like rain, salmon pervades the identity of Seattle, producing a hybrid conception of the city as defined by and through nature.

The Allure of Nature in the Pacific Northwest and Seattle

At the beginning of the nineteenth century, the Pacific Northwest was a place of geographic isolation; indeed, the region has had appreciable Anglo settlement for only the past century and a half.[29] San Francisco became the closest link to civilization; the 1849 gold rush and Pacific Northwest towns served as principal trading outposts throughout the remainder of the 1800s. Settlement of the region began after the Civil War but continued to be relatively unpopulated until the 1880s and 1890s, when settlers came en masse following the 1883 completion of the Great Northern Railroad.[30] Railroad marketing campaigns characterized the region as the Great Northwest and the Great Pacific Northwest to lure settlers and vacationers with the romance of experiencing the last region to be settled in the continental United States.[31]

A significant element in marketing the Pacific Northwest to newcomers was its intrinsic natural beauty. The region was spared the ravages of the Civil War and rapid industrialization that decimated the landscape in other parts of the country; Raban argues that Anglo populations began to settle the Pacific Northwest just as the notion of the Romantic Sublime came to dominate environmental thought.[32] The region was an ideal place to conceive of how the landscape *should* look, with iconic features reminiscent of the Swiss Alps, the German Forest, and the English Lake District. Raban writes, "It was a quickly established convention that Northwest water was sufficiently still to hold a faithful reflection of a mountain for hours at a time" and historian Robert Bunting adds that "if America was 'Nature's Nation,' no place looked upon itself more self-consciously as 'Nature's Region' than the Northwest."[33]

With respect to Seattle, early twentieth-century historian Welford Beaton sums up the allure of the city to early residents as follows:

We who live here are persuaded that nowhere else on earth is a city favored such as ours. With every facility for great commerce by land and sea, we combine an aesthetic perfection that no other commercial center on the globe can match. Over hills and across valleys the city stretches, and from every doorstep there is a view of mountain and water. Roses, which pay scant attention to the calendar, climb over the palace of the millionaire and the cottage of the artisan. From behind the Cascade Mountains the sun comes up each day and at night falls beyond the jagged peaks of the Olympics, his last rays lighting up a golden path across Puget Sound to the shores that Seattle rests upon. To the south the vista holds Rainier's hoary peak rising majestically above all other heights in any of our states. To the north is the peaceful pathway of water that brings the fleets of all the world to lay commerce on Seattle docks. The climate is a peaceful one, given to no excesses and but scant indulgence in snow or frost. All the year around our lawns are green.[34]

The innate beauty of Seattle and the Pacific Northwest was not to be admired from a distance. The lush, green landscape was understood to be full of inexhaustible natural resources ripe for extraction by industrious settlers. As such, nature served double duty, first as an aesthetic amenity for pleasure and wonder, and second as a source of employment and profit.[35] Early fur-trapping activities gave way to timber and fishing industries, and these economic activities attracted a population that was largely Anglo and male. The demographics of the region at times fueled an environmental determinist argument familiar to Northern Europeans and Scandinavians that promoted the climate as ideal for Anglos to prosper intellectually and spiritually.[36] Today, the regional demographics continue to be skewed toward Anglos, with the cities of the Pacific Northwest having some of the lowest populations of African Americans in the United States, although there are significant numbers of Asian Americans and Latinos.[37]

Although the population of the Pacific Northwest tends to be relatively homogenous, the region is heavily influenced by Asian and European cultures as well as the region's original inhabitants, the various Native American tribes.[38] Reflecting on the importance of Native American culture to the region, Raban argues that "its Indian . . . past lies very close to the surface, and native American conceptions of landscape and land use remain live political issues here."[39] The Native American heritage of Seattle has a strong influence on its international reputation, particularly the city's namesake, Chief Seattle or Sealth. Chief Seattle was the leader of the Suquamish and Duwamish tribes and is known throughout the world for his famous speech from 1854, a part of which reads:

The earth does not belong to man; man belongs to the earth.
This we know.
All things are connected like the blood which unites one family.
All things are connected.
Whatever befalls the earth befalls the sons of the earth.
Man did not weave the web of life; he is merely a strand in it.
Whatever he does to the web, he does to himself.[40]

The speech of Chief Seattle glorifies the harmonious relationship between Native Americans and nature and reflects an ecological ethic that seems to be lacking in industrialized societies. A significant problem with this story is that Chief Seattle never uttered these influential words; the "Chief Seattle Speech" is a product of American screenwriter Ted Perry, who was hired by the Southern Baptist Radio and Television Commission in the 1970s to draft a script for an environmental documentary. Perry's words reflect the 1970s agenda of the Club of Rome environmental group rather than the beliefs of late nineteenth-century Native Americans of the Pacific Northwest.[41] Yet the myth of Chief Seattle continues to inform the environmental rhetoric of the region, defining a landscape ethic supposedly rooted in local Native American philosophy. Furthermore, it is well known that Native American populations, including Chief Seattle's tribes, drastically altered the landscape of Seattle and the Pacific Northwest region, particularly through fire management techniques.[42] Early Anglo visitors were struck not by the Native American reverence for the land but rather by the amount of damage they inflicted upon it.[43] This observation is not meant to suggest that the landscape alterations of Native Americans were equivalent to subsequent activities by Anglos in the Pacific Northwest but instead to acknowledge that the landscape encountered by the first Anglo settlers was far from virgin.

Geographic narratives are particularly important for framing the benefits and drawbacks of particular places while promoting normative ideals for how residents *should* relate to their material surroundings.[44] The ideology of Seattle as a place where humans and nature can coexist in harmony is one that continues to this day. One need only look to the homegrown outdoor recreation companies such as REI and Eddie Bauer; the popularity of outdoor pursuits such as hiking, camping, and skiing; the aforementioned Metronatural campaign of the Seattle Convention and Visitors Bureau; or any tourist guidebook to recognize that nature pervades and defines the city. This reputation is due not only to the physical landscape but also to the various stories that have been told and retold by residents, urban boosters, tourism proponents, and visitors since the city's founding. Nye argues that "people tell stories in order to make sense of their world, and some of the most frequently repeated narratives contain a society's basic assumptions about its relationship to the environment."[45] The stories that Seattleites told and continue to tell about nature are part and parcel of Seattle's standing as a city in the Promised Land.

Improving the Promised Land

The logging and milling of "big nature" in the land surrounding Seattle created a small industrial town by the late nineteenth century.[46] However, urban boosters recognized that the growth of the settlement could not rely on its original economy of natural resource extraction and would need to expand to compete with the larger towns of Tacoma, Portland, and San Francisco. To spur economic growth, a new interpretation of nature would emerge to frame Seattle's nature as beautiful but also incomplete. Historian Coll Thrush summarizes this new perspective of a town in dire need of human improvement: "Seattle was a bad place to build a city. Steep sand slopes crumbled atop slippery clay; a river wound through its wide, marshy estuary and bled out onto expansive tidal flats; kettle lakes and cranberried peat bogs recalled the retreat of the great ice sheets; unpredictable creeks plunged into deep ravines—all among the seven (or, depending on whom you ask, nine or fifteen) hills sandwiched between the vast, deep waters of Puget Sound and of Lake Washington."[47]

Beaton makes a similar observation, stating that the glaciers that shaped the landscape left the city as a "tousled, unmade bed" with undulating hills that provided breathtaking views but hindered economic development.[48] This new perspective on nature in Seattle at the turn of the century recognized the landscape as both a blessing and a curse; the city was notable not only for what it contained but also for what it lacked.[49] Furthermore, some of the most rigorous promoters of economic development held an implicit belief that nature's abundance and the moderate climate led to unproductive behavior.[50] Settlers could achieve extraordinary results with little effort, so there was a lack of impetus for urban development.[51] The curse of the bountiful landscape was that it led to idleness and a lack of ambition.[52] To correct the problems of the city, the landscape would be subjected to rigorous transformation so that nature and its residents could realize their full economic and social potential. The Prometheans of Seattle recognized that the region had been blessed with a surplus of natural resources that could and should be exploited to achieve widespread societal progress. Subsequent processes of "improvement" would require significant human intervention to change the human/nature relationship through the application of the latest technological advances, resulting in the creation of the contemporary municipal government and the economic engine of the city.

Three events in the last decade of the nineteenth century fueled Seattle's transformation from a fledgling settler outpost into a contemporary city defined by urbanization, large-scale capital investment, and the bureaucratization of government and business.[53] First, the aforementioned arrival of the Great Northern Railroad in 1893 allowed Seattle to compete economically with its closest urban rival, Tacoma. Second, the discovery of gold in the Klondike territory of Canada in 1897 transformed Seattle into a major trading center, with the 1880 population of 3,500 residents quickly growing to over 240,000 by 1910.[54] Third, an event of large-scale destruction allowed the city to rebuild from scratch; in June 1889, the entire downtown burned to the ground. The fire consumed fifty blocks and resulting in property damages estimated at $10 million, although miraculously no human lives were lost.[55]

The 1889 fire highlighted the need for a consistent water supply that could provide fire protection and support the growing urban population. Thus, the development of water supply infrastructure had multiple implications for residents of the city, not only as a means to protect the past and present city but also to ensure a prosperous future. It would involve the transformation of natural resource flows into capital flows, a common formula for civic improvement across the United States at the time.[56] From a public works perspective, the fire was a godsend because it created a tabula rasa upon which new infrastructure networks could be built using an emerging rational and systematic planning logic; it also fueled widespread civic spirit directed at rebuilding the decimated downtown.[57] Beaton writes, "While the ruins were yet smoldering the people of the stricken city met in the Armory to plan the rehabilitation of Seattle."[58] Rehabilitation was completed in a few short years with the rebuilding of the original wood buildings in brick and stone. More important, the rebuilding of Seattle would involve several massive engineering projects that would remake the entire landscape and "improve" nature in Seattle.

Bringing Water to Seattle

Before the 1889 fire, residents of Seattle relied on pumps from Lake Washington for their water supply, but the pumping capacity proved inadequate for fire protection. The first inclination by city officials after the fire was to increase the capacity of the existing system, but this strategy was quickly abandoned in 1892, when Reginald H. Thomson was appointed as city engineer (see figure 5.2). Thomson, a self-trained engineer from Indiana, worked as a surveyor before he was hired by the City of Seattle to head up its nascent engineering department. He argued that Lake Washington was inadequate for the future growth of the city because of the expense required to pump the water to the city's residents; he called instead for a gravity-fed water supply from the Cedar River, a waterway that originates in the Cascade Mountains and flows into the southern end of Lake Washington.

Figure 5.2
Seattle city engineer R. H. Thomson, circa 1931. *Source*: MOHAI 1986.5.43455.

Seattle voters approved a $1 million bond to build twenty-two miles of pipe that would create the Cedar River water system, and it was opened in January 1901 (see figure 5.3).[59] The municipality subsequently used the consistent, clean water supply to annex neighboring cities such as West Seattle and Ballard; today, Seattle acts as a wholesaler of Cedar River water to several neighboring municipalities, including Bellevue, Kirkland, and Redmond.[60] The new water supply would expand the "ecological frontier" of the city while cementing Seattle's status as the dominant city in the region.[61]

Thomson was undoubtedly one of the most important individuals in the growth of Seattle. He served as city engineer under seven mayors from 1892 to 1911, and the Cedar River water supply system would be only the first of many major projects he undertook to transform Seattle into a contemporary city. Thomson was a proponent of big civic improvement projects, and—like other Progressive urban reformers—he believed that social problems could be solved by more and better engineering.[62] In this way, he was the quintessential technocrat, using his engineering vision to

Figure 5.3
Construction of the Cedar River water supply in 1899. *Source*: University of Washington Libraries, Special Collections, UW 27484.

conflate technical and social progress while purportedly staying above the fray of partisan municipal politics. His work on Seattle's physical landscape would perfect nature's designs and, in doing so, improve society, much like Colonel George Waring Jr., the vaunted nineteenth-century U.S. sewer expert, and William Mulholland, the famed municipal engineer who brought water from the Owens Valley in eastern California to Los Angeles.[63] At the turn of the twentieth century, Thomson's most pressing engineering problems for Seattle included water supply, sewerage, and transportation. And he recognized that they were all related by the common element of water.

With respect to sewers, the first formal system was initiated in the city in 1875 with wood troughs and boxes, while the use of more durable materials began in 1883.[64] Three years before Thomson would become city engineer, the municipality hired Waring to produce a report for the city on sewerage. Waring advised the municipal government to build a sewer system for sanitary wastes and create a surface drainage system for stormwater, but municipal officials balked at the cost and design of

Table 5.1
Total miles of sanitary sewer in Seattle, 1891–1950

Year	Total miles	Year	Total miles
1891	14.9	1930	802.10
1900	60.45	1940	863.15
1908	212.32	1950	988.09
1924	628.63		

Source: Brown and Caldwell 1958.

his proposed separated sewer plan. They solicited a second sewer design from Chicago engineer Benezette Williams, who recommended a system with much larger capacity but also combined the sanitary and stormwater volumes in a single pipe. The municipality approved Williams's combined sewer design and construction on the system began in 1890.[65]

After being appointed as city engineer, Thomson assessed the Williams plan and largely adopted it but made one significant change: instead of discharging untreated sewage volumes into Lake Union and Lake Washington, he called for deep outfalls in Puget Sound. He felt that the lakes had insufficient capacity to absorb the pollutant loads from the growing city, a far-sighted observation at the time. Under his guidance, the city built two massive trunk sewers in the 1890s and went on to build a comprehensive combined sewer system that is still used today in the central core of the city (see table 5.1).[66] However, Thomson's embrace of Williams's combined sewer design would be a costly mistake, due to future requirements for wastewater treatment as well as significant increases in stormwater volumes from impervious surfaces in the city. Like all infrastructure champions of this time period, he could not anticipate the changing material and cultural conditions of the city that would create unintended problems in the future.

Engineering the Landscape for Economic Development

Designing and constructing water and sewer systems were enormous endeavors at the turn of the twentieth century and required massive amounts of funding as well as the mastery of engineering techniques that were changing rapidly.[67] However, the building of Seattle's water and sewer networks would pale in comparison to Thomson's most famous engineering project. Surveying the landscape of the city, he noted a preponderance of hills and recognized them as an impediment to growth. In his memoirs, he famously remarked, "Looking at [the] local surroundings, I felt that Seattle was in a pit, that to get anywhere we would be compelled to climb out if we could."[68] This perspective mirrored that of other Progressives who interpreted Seattle's landscape as a hindrance to economic growth. Thomson's proposed solution for bringing Seattle out of its pit would involve a series of massive projects to erase the hills of downtown Seattle and flatten the grades of its transportation routes. The regrading projects would correct the impediments to urban growth dictated by the original landscape and create a level platform on which commerce could flourish, a common goal for nineteenth- and twentieth-century Promethean actors. Thomson's vision for the city was also a topographic one: economic development could occur only on a rational and abstract platform rather than the heterogeneous, ever-changing landscape that currently existed.

The first regrading project began in 1898 on Denny Hill to the north of the downtown and used steam shovels and carts to remove soil. Progress on removing the hill was slow, but this would change with the arrival of Cedar River water in 1901, when manual excavation was replaced by hydraulic sluicing techniques developed for gold mining operations on the Pacific Coast. Hydraulic nozzles called *giants* were modified with a ball-and-socket joint to allow for horizontal and vertical movement, and large nozzles 2.5 to 3.5 inches in diameter were used to create water pressures of 75 to 100 pounds per square inch (see figure 5.4). The adoption of water as the medium of excavation significantly accelerated the regrading efforts by creating an industrial process of landscape alteration.[69] Beaton provides a vivid description of the regrading process: "Day and night great streams of water incessantly ate their way into the yielding hills, steam shovels chewed immense holes and spit their mouthfuls into waiting wagons, houses stood dizzily on freshly created peaks, or moved out of the way of the devouring progress to return later and lower themselves to the new levels that were provided."[70]

Massive amounts of water were required to remove the hills and grades in downtown Seattle, the majority coming from the Cedar River water supply and supplemented by a pumping plant on Elliott Bay. For example, on the Jackson Street regrade on the south side of the central business district, the giants consumed between 9 and 12 million gallons of water per day, almost a third of the total Cedar River water supply. The Cedar River would also supply kinetic energy to power the giants and hydroelectric power for electric lights so that sluicing operations could continue 24 hours a day.[71] Had it not been for the forgiving geologic conditions of the hills—composed of sand, gravel, clay, and hardpan as opposed to

Figure 5.4
Hydraulic sluicing activities at Bell Street and Fourth Avenue in downtown Seattle. *Source*:
University of Washington Libraries, Special Collections, Lee 20022.

rock—the regrading efforts would have been impossible.[72] In short, available technology as well as geologic conditions in Seattle conspired with the Progressive logic of landscape improvement to create a wholly new relationship between Seattleites and their surroundings.

The hydraulic sluicing technique greatly accelerated the regrading efforts and was used for the majority of the almost sixty regrading projects over the next three decades (see figure 5.5). The elevations of more than twenty downtown streets were altered, and several large hills were removed with the financial support of downtown property owners, who would benefit from increased property values and business revenues (see table 5.2).[73] In some cases, the municipal government was forced to use the power of eminent domain to compel uncooperative residents to support the regrading efforts. Some of the most indelible photographs of Seattle from the early twentieth century feature "spite mounds" with buildings resting on precariously small patches of land spared from the giants.[74] All of the recalcitrant property owners would eventually give in and the

Figure 5.5
The regrading of Second Avenue in 1906, with the Washington Hotel towering above the
surrounding downtown blocks. *Source*: MOHAI 2002.3.383.

Table 5.2
A sample of statistics from Seattle regrading projects

Location	Maximum cut depth (feet)	Soil removed (million cu. yd.)
Denny Hill	110	5.4
Jackson Street	84	3.4
Dearborn Street	108	1.6

Source: Dimcock 1928; Dorpat and McCoy 1998.

mounds would be erased, fulfilling Thomson's master plan of creating a rationalized landscape for economic growth. The regrading projects are the embodiment of Promethean hubris working to perfect nature through technology, as reflected in Beaton's sympathetic description of the projects: "The hills raised themselves in the paths that commerce wished to take. And then man stepped in, completed the work which Nature left undone, smoothed the burrows and allowed commerce to pour unhampered in its natural channels."[75]

The regrading activities in Seattle can be understood as part of a long history of reworking the landscape for human ends, a practice that would reach its zenith with experiments in "geographical engineering" in the 1950s and 1960s with the moving of vast quantities of earth and rock via nuclear explosions.[76] Estimates on the total volume of soil moved during the regrade projects vary, with some figures as high as 50 million cubic yards or about one one-eighth of the Panama Canal excavation, the measuring stick for large engineering projects at the time.[77] The regrading projects illustrate the tight connection between urban development, landscape alteration, and technological progress.[78] An irony of the regrade projects is that improved building and transportation technologies were introduced while the regrading projects were underway, making the reduced grades in downtown Seattle unnecessary. Specifically, the emergence of the automobile and more durable paving materials meant that the previously impassable hills could now be easily navigated.[79] The regrades were seen as ultimate, permanent solutions to engineer the city, but they could not keep up with continually evolving technological, environmental, and social conditions.

Rationalizing the Boundary between Land and Water

Removing the hills of Seattle had multiple consequences. Thomson sold the regrading projects as a means to unite the city's divided neighborhoods by erasing the vertical relief that constrained the free flow of people and goods, but the regrades were also central to rationalizing the shifting boundary between water and land. Raban notes that "Seattle was built on pilings over the sea, and at high tide the whole city seemed to come afloat like a ship lifting free from a mud berth and swaying in its chains."[80] The waterfront consisted of a shifting amalgam of water and land that resisted the creation of a permanent foundation for economic growth. Seattle's economic future was compromised not only by a hilly landscape but also one that was impermanent, with the boundary between dry land and water being continually renegotiated by tidal activity. This relationship was particularly evident in the area south of the downtown region, where the Duwamish River emptied into Elliott Bay. The tideflats were a desirable location for economic expansion of the central business district because of the ready access to deep water for transportation of goods, but this location was hindered by the presence of thousands of acres of estuary lands (see figure 5.6).[81] From the perspective of Promethean actors,

Figure 5.6
The tideflats and waterfront south of downtown Seattle as seen from Beacon Hill circa 1898. *Source*: MOHAI 1983.10.6049.4.

the estuary was a disordered landscape that introduced uncertainty and constant change rather than solidity and stability. It was an unfinished landscape in dire need of improvement.

Filling of the tidelands began informally in the nineteenth century, as volumes of municipal solid waste and sawdust from the early sawmills were deposited on these unbuildable lands, mirroring efforts throughout the United States to rationalize these landscapes for productive use.[82] The regrading efforts produced massive quantities of fill for the low areas of the city, and the industrial process of regrading was extended to involve the systematic filling of the tidelands. Spoils from the regrading activities were washed into sluiceboxes and loaded into carts for transportation to the tidelands and low areas, where they were dumped.[83] The goal was to establish grades two feet above extreme high tide; in some areas, this meant raising tideflat elevations as much as forty feet. Overall, an estimated twelve hundred acres of tideflats were filled with spoils from the regrading projects.[84]

The tideflat reclamation project was supplemented in 1914 with dredging activities in the lower Duwamish River to straighten and deepen the channel as well as fill in the estuaries at the mouth of the river. Within eight years, contractors had moved 24 million cubic yards of soil and replaced the river delta with what was at the time the world's largest artificial island, Harbor Island, which would eventually become a platform for industrial activities.[85] In the mid-1930s, the City of Seattle received federal funding from the Works Progress Administration to build the Alaskan Way seawall along the downtown waterfront. The seawall created a permanent boundary between land and water, ensuring that waterfront buildings in the central business district would not be undermined by water. These activities created "a new place for nature in the city subjected to an ever-greater disciplinary control" and formalized the divide between nature and culture.[86]

Big Engineering and Social Inequity

With all of his large engineering projects, Thomson espoused social and economic reform through the modification of material conditions. Historian Matthew Klingle notes that "clean water and level land were ethical as well as economical and political goals for Thomson."[87] The material reordering of the city would simultaneously fulfill the Promethean goal of controlling nature while ensuring societal progress: the classic formula of Progressive urban reform.[88] However, the grading and filling activities were not nearly as benign as Thomson claimed, due to the geographic distribution of Seattle's population. Middle- and upper-class residents in Seattle lived on the ridges and hills, looking down upon the polluted lowlands below that were occupied by low-income, minority, and transient populations.[89] The regrading and filling projects raised the lowlands and created more land for economic growth, and as a consequence, populations who lived in these lowlands were pushed further afield and concentrated in areas outside of the downtown business district.[90] As noted earlier, the regrading activities resulted in the relocation of some middle- and upper-income residents on the ridges and hills, but this relocation paled in comparison to the filling of the tidelands because the latter activities encompassed a larger area and were more comprehensive in nature. As such, the Promethean activities in Seattle are an example of "how nature and society combine in the production of socio-spatial fabric that privileges some and excludes many."[91]

During the Great Depression, shantytowns sprung up on the waterfront and the newly filled tidelands south of downtown. In 1935, an estimated four to five thousand residents lived on the tidelands in an informal "Hooverville" community (figure 5.7). In 1942, municipal officials cited national security issues related to waterfront activities for the war effort as an excuse to burn Hooverville to the ground.[92] Fire would once again be used to reorder the landscape, but this time it was a deliberate intervention aimed at improving the social order. Meanwhile, Progressive idealism espousing the civic virtue of massive engineering efforts hid the removal of unwanted residents, filth, and blight from the city. Those living at the bottom of society, both economically and physically, shouldered the negative consequences of rationalizing the Seattle landscape for economic gain.

In 1931, the regrade projects came to an unceremonious end due to a gradual erosion of public support as well as the arrival of the Great Depression.[93] Thomson was an adept political strategist during his tenure as city engineer, but his regrading efforts had turned Seattle into "one vast reclamation project" for three decades, and residents were fed up with living and working in a perennial construction zone.[94] The disenfranchised and those who were displaced or whose property values were diminished by the regrading efforts finally succeeded in fighting off the municipal government. Reflecting on this period in Seattle's development, Klingle notes, "The genius of the engineer was to make the ceaseless destruction and rebirth of the city an exercise in progress, but progress had its limits. Even Thomson could not annihilate all of Seattle's hills at once."[95] Despite the end of the regrades, the newly created landscape was a permanent

Figure 5.7
The informal settlement known as Hooverville on the filled tidelands south of downtown Seattle in March 1933. *Source*: MOHAI 1983.10.10788.

one and continues to serve as the foundation for the city. Local historian Murray Morgan notes, "The engineers made it possible for a modern metropolis to be built on the half-drowned mountain that lies between Puget Sound and Lake Washington."[96] Surprisingly, no major infrastructure project or civic building bears Thomson's name despite his central role in the building of Seattle. In the 1950s, the municipal government proposed the R. H. Thomson Freeway to connect Renton to University Village in Central Seattle, but the plan was eventually defeated in the early 1970s by antihighway and environmental activists.[97] Thomson remains the unacknowledged "hero" in the creation of contemporary Seattle.

The Land Refuses to Be Tamed

An irony of the transformation of the landscape in Seattle was that Thomson and the municipal government marketed these efforts as a means to create a stable foundation for the city while unintentionally creating new instances of perennial instability. Klingle notes that Seattle's regrading experts had created "hybrid landscapes, neither fully natural nor fully under human control. Engineers had produced, very often, landscapes more dangerous and less reliable than those they had altered."[98] Prometheans were committed to the impossible goal of complete control over nature while their activities often shifted problems either geographically or temporally rather than resolve them.[99]

From the earliest times, landslides and unstable soil conditions plagued Seattle residents; in many cases, civic improvement projects served to exacerbate these conditions. The glacial history of the region resulted in the deposit of unconsolidated or partially consolidated soils; although most of these sediments are compacted under several thousand feet of glacial ice, they are not solid rock. As noted earlier, this geologic configuration was beneficial for hydraulic sluicing activities but served as a less than desirable foundation for urban development, particularly on steep slopes.[100] Furthermore, the subsurface consists of a lattice of water flows that slowly loosens the layers of sediment and eventually causes landslides. The construction of roads, buildings, and underground utilities starting at the end of the nineteenth century would change these subsurface drainage patterns, resulting in landslides in unanticipated areas. Alteration of the landscape for urbanization not only implicated surface features but also the complex amalgam of soil, rock, and water in the subsurface. Landslide mitigation emerged as another municipal service to mediate the relationship between Seattleites and the landscape (see table 5.3).

Table 5.3
City of Seattle slide removal activities, total cubic yards of material removed, 1923–1935

Year	Hand removal	Gas shovel	Year	Hand removal	Gas shovel
1923	5,316	18,360	1930	1,222	2,724
1924	1,684	20,584	1931	2,614	*
1925	2,094	31,365	1932	2,523	*
1926	832	6,496	1933	5,754	*
1927	3,365	5,972	1934	41,763	*
1928	2,317	32,886	1935	19,250	*
1929	1,032	12,258			

Source: City of Seattle Streets and Sewers Department Annual Reports, 1924–1936.
* = No data reported.

To address the unstable geology of the city, the federal Works Progress Administration provided funding to the City of Seattle's Engineering Department in the mid-1930s to complete twenty-nine drainage projects. These projects involved the construction of trenches, tunnels, footings, and retaining walls to redirect subsurface water flows and reduce landslide incidents.[101] As with other New Deal programs, the drainage projects were a combination of infrastructure improvement and employment for local residents; a stipulation of the WPA funding required the municipality to use unskilled labor and hand tools. Between 1935 and 1941, more than seven hundred laborers dug an estimated twenty thousand feet of drainage trenches at a cost of $1 million (see figure 5.8).[102] Thomson's Progressive logic of improving society by improving nature was extended during the Great Depression through additional public infrastructure projects. The drainage improvements provided by the WPA funding were not as newsworthy as other infrastructure projects in the Pacific Northwest, such as the building of the Grand Coulee Dam, but they would serve to further "invent" the landscape that exists in Seattle today.

The drains continue to function today, and municipal staff members check them periodically to ensure that they are functioning as designed. A municipal staff member reflecting on the WPA-funded work in the 1930s notes, "God bless the WPA for putting in those drains. But they didn't map

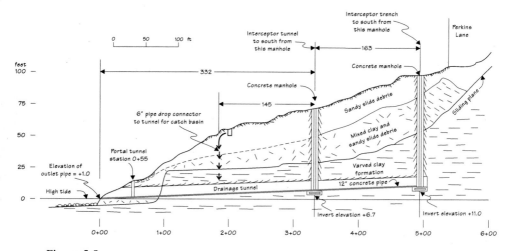

Figure 5.8
Cross-section of a WPA slide mitigation project in the Magnolia neighborhood of Seattle with two vertical interceptor tunnels and a horizontal drainage tunnel, all dug by municipal laborers with hand tools. *Source*: Evans 1994.

them very well and it's a little unclear who is responsible for them."[103] In other words, the drains continue to be crucial to maintaining habitable conditions aboveground, but their subterranean character makes them difficult to manage. A major storm in 1996 resulted in three hundred landslides and caused damage to municipal property in excess of $30 million as well as an undetermined amount of private property damage. The municipal government subsequently created a comprehensive strategy to address landslide issues.[104] A contemporary map of landslide and liquefaction zones in Seattle shows a close correlation between the land-water interface and the shifting landscape (see figure 5.9). The Promethean Project did not separate the city from nature but rather created new dependencies that required continuous management. As Kaika notes, "modernization is an ongoing project in which natures, cities, and people are woven together in an inseparable dialectic of creation and destruction."[105]

The Unanticipated Consequences of Economic Growth

The big engineering projects undertaken since the late nineteenth century in Seattle were intended to improve upon nature and create a city that could prosper while enjoying the advantages of a beautiful natural setting. Instead, the municipal government created a hybrid ecotechnical landscape that was at once controlled and out of control, predictable and unpredictable.[106] But the seemingly improved landscape created by Thomson and the municipal government would lead to unprecedented population growth and urban expansion in subsequent decades. Seattle's population increased steadily from the turn of the century, starting with fewer than one hundred thousand residents in 1900 and growing to over five hundred thousand residents by 1960 (see figure 5.10).

The tipping point for population growth in Seattle and the surrounding areas would come in the 1940s, as water quality in Lake Washington began to change visibly due to sewage inputs from Seattle and the surrounding suburbs. The lake, the second largest in the state with a length of twenty-two miles and a width varying from one to four miles, had also undergone "improvement" as Seattle grew.[107] While the Duwamish industrial channel was being created in 1914 to the south of the central business district, similar construction projects were underway to remold the hydrologic flows north of the city. In 1916, the municipal government completed a ship canal to connect Puget Sound to Lake Union and Lake Washington via a series of locks.[108] Completion of the canal realized a long-held dream by the municipal government and business interests to

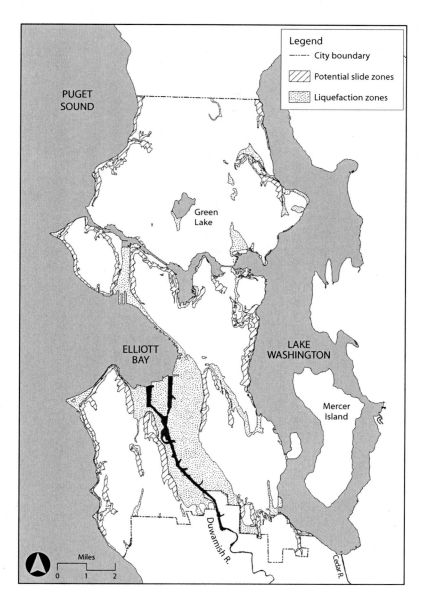

Figure 5.9
Potential slide and liquefaction zones in Seattle today.

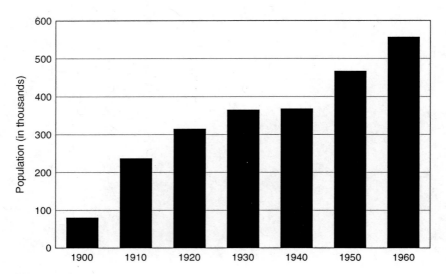

Figure 5.10
Population growth of Seattle, 1900–1960. *Source*: City of Seattle 2008.

connect the waterbodies and thereby open them to commerce and recreation. The ship canal dropped the elevation of the lake by an average of ten feet, creating new shoreline for residential development, and redirected the lake outflow through the ship canal to the west rather than the Duwamish River to the south.

Thomson made a wise decision in the late nineteenth century to direct major sewage outfalls away from Lake Union and Lake Washington, but nonetheless, smaller sewage outfalls from Seattle and other municipalities continued to be directed to the water bodies. By 1922, Lake Washington was receiving thirty raw sewage outfalls from the east side of the city, and residents began to recognize significant impacts on the water quality of the lake.[109] The city built three rudimentary wastewater treatment plants in 1924 at a cost of $2.5 million but later abandoned these plants in favor of an interceptor sewer, constructed in 1936, to divert all Seattle sewage outfalls from the lake.[110] Despite these efforts, growth of the surrounding suburbs, particularly after World War II, would lead to further deterioration of the lake's water quality. The Mercer Island floating bridge was opened in 1940 and provided a convenient automobile connection between Seattle and the eastside. Urban sprawl around the lake created twenty-one independent sewer districts by 1955, and these communities were dumping both treated and untreated waste into the lake. Visibility in the lake was over nine feet in 1938 but dropped to less than three feet by the early 1950s.[111] Thick mats of algae began to appear, and public health officials had to shut down swimming areas due to health risks. The notion of "eutrophication" entered the public lexicon as scientists at the University of Washington began to characterize how the influx of nutrients from untreated sewage volumes in the surrounding cities to the lake resulted in an overly enriched ecological system.[112] The deteriorating water quality of Lake Washington became a highly visible symptom of the unintended effects of urban growth in the region. Findlay writes: "The region's identifications with its natural resources—nurtured for so long by railroads and other boosters—came to have new implications: calls for preservation rather than exploitation of those resources became stronger. Continued growth, especially unmanaged growth, came into question. Overcrowding and pollution threatened the good life that was the trademark of the Pacific Northwest for so long."[113]

It was during this time—a decade before environmental pollution would emerge as an issue of national importance—that the Promethean approach would be discredited in Seattle. There was widespread acknowledgment that the dilemma of nature had not been solved by engineering improvements. Kaika describes this downfall of the Promethean Project: "From tamed and controlled (the prerequisite for development), nature is now the source of crisis, a potential impediment to further development."[114] The rationalization of the landscape in the early twentieth century had the unanticipated effect of massive population growth that created undesirable environmental pollution. But unlike the social inequity and land instability issues described previously, the deterioration of Lake Washington was a highly visible problem that would spur residents and the government into action to redefine the relationship between Seattleites and nature.

A Regional Approach to Protect Water Quality

Beginning in the 1950s, Lake Washington became a living laboratory for University of Washington scientists to determine the causes and effects of adding nutrients to a large water body and to devise strategies to reverse the impacts of pollution. This "pioneer whole lake experiment" was an unprecedented synergy of scientific and public awareness of an environmental problem at the time.[115] Political activists in Seattle recognized that water pollution was tied to larger problems of urban growth that would require a coordinated, large-scale effort to resolve. Just as there was a concerted effort in the first half of the twentieth century to tame nature, in the second half there was an equally concerted effort to save nature from the unintended consequences of urban growth, signaling a shift from the second stage of the Promethean Project (widespread improvement) to the third stage (crisis).

In the early 1950s, an urban reformer named Jim Ellis initiated a proposal to solve Lake Washington's pollution by forming the Municipality of Metropolitan Seattle, or Metro.[116] Metro would manage growth through jurisdiction over water supply, sewerage, solid waste management, transportation and land use planning, and park administration.[117] Inspired by Toronto's example of regional governance, the "super utility" would have taxation power to fuse the old faith in engineering with new public interests in ecological protection and environmentalism.[118] Ellis noted, "Seattle is a young city with a chance to lick its metropolitan problems before being swallowed up by them."[119] Thomson had created an urban growth machine for Seattle but failed to account for the consequences of population growth in the areas adjacent to but outside of the municipality's jurisdiction. The economic gains afforded by the rationalized landscape of the city had extended population growth and land development beyond the regulatory reach of the city. Proponents of regional governance in the 1950s were attempting to reconcile the economic and environmental landscapes of the region through the comprehensive provision of public services.

Not surprisingly, the proposal for a strong regional government was resisted by many residents, particularly those in Seattle's suburbs who feared the consolidation of political power in the city as well as the threat of higher taxes and restrictions on property rights.[120] These residents saw themselves as outside the network of economic growth and environmental protection that Seattle had fostered throughout its history. The difference in perspective between Seattleites and other regional residents was evident in the first regional vote to authorize the formation of Metro in March 1958. Seattle voters approved the ballot measure, but residents in smaller cities and suburbs soundly rejected it. Undeterred, Ellis and his group of regional governance proponents revised the authority's mandate, significantly limiting Metro's jurisdiction to only sewage treatment and excluding those districts (particularly in south King County) that had rejected the first proposal. Their revised strategy was a success; Metro received voter authorization by a majority of Seattle and suburban voters in September 1958.

After forming, Metro moved quickly, and in 1961 proposed a ten-year sewage system plan that would comprise 231 square miles at a cost of

$125 million. The plan, funded through a $2-per-household monthly sewer charge, involved the construction of an interceptor sewer around the entire perimeter of Lake Washington as well as facilities to treat waste volumes before discharge into Puget Sound. At the time, this was the most expensive sewage diversion project in the nation.[121] Metro also entered into fifty-year contracts with member cities and neighboring sewer agencies, and by 1962, had signed contracts with fifteen municipal sewage districts representing over three hundred thousand customers.[122]

Diversion of sewage discharges began in 1963, and water quality conditions in the lake improved almost immediately. Visibility in 1967 was 2.5 feet, 9 feet the following year, and 20 feet by 1977—twice the visibility as measured in 1938. The diversion process was completed in February 1968, and Metro received international recognition for one of the most successful lake cleanup efforts ever undertaken.[123] The effort was similar to the large engineering projects undertaken in the first half of the century to replumb and regrade the city, but now the equation was reversed: big engineering was needed to improve nature by controlling the impacts of humans.[124] Furthermore, the work on Lake Washington added environmental protection to the Promethean Project of technological development. A change in societal values in response to post–World War II urbanization was translated into an amended mission for technical experts who governed the environmental flows of the region. The call for environmental protection did not threaten the technomanagerial elite but rather strengthened their position while extending their geographical reach as environmental managers. Nonexperts were involved in exposing the problems of pollution, but it was the experts who were charged with designing and implementing solutions that could be carried out by local and regional governments.

Despite the claims of Metro and its allies, the replumbing of the region to save Lake Washington was not a win-win solution for all residents in the region. Saving the lake came at the expense of the waterbody receiving the diverted sewage, the battered Duwamish River. Already burdened by dredging and straightening activities to accommodate industrial development in the early twentieth century, the fishermen and Native Americans who relied on the Duwamish for recreation and sustenance were further affected by the Metro interceptor sewers and treatment plants that defined the river as the ultimate drain of the region.[125] As a result, those who lived closest to Lake Washington received the greatest benefits, while those who lived in the bottomlands of the city were once again burdened with the negative side effects of economic and population growth. Klingle notes,

"Just like the Progressive agenda, Metro's agenda benefited those with power."[126] The triumph of cleaning up Lake Washington involved shifting environmental risk and aesthetically undesirable conditions to a different locale, the Duwamish River, with particularly detrimental impacts on low-income populations who lacked political power to protest this strategy.

Beyond the inequitable conditions created by Metro's sewage diversion projects, the regional agency failed in its original mission to control and direct growth. By limiting its authority to sewage treatment, Metro promoted a "technical fix" to resolve the regional growth problem by creating a mandate for universal sewage treatment and, ironically, facilitated more growth in the distant suburbs that could rely on the new interceptor sewers to serve new community development. Klingle argues that "Metro's sewers spared the lake but did not rein in unbridled growth. They merely pushed the waste elsewhere and let the building continue."[127] The implicit message of Metro's approach was that environmental problems could be overcome with a comprehensive regional infrastructure network. Meanwhile, the region's residents continued to be wary of regional planning and governance, as evidenced by voter rejection of several efforts by Jim Ellis and Metro supporters to expand their authority to mass transit, park management, and other public services in the 1960s and 1970s.[128]

Conclusions

The history of Seattle's development from its origins in the nineteenth century to the 1960s suggests that the city is not as close to nature as its contemporary Metronatural reputation suggests. Time and local narratives about the "naturalness" of the city have hidden the massive engineering projects of Thomson and the municipal government that transformed Seattle from an organic city of the nineteenth century into the rational, planned city of the twentieth century. Furthermore, the creation of this hybrid landscape centered on one overarching goal: economic growth. Progressive proponents at the turn of the century justified the big engineering projects as a way to improve society by improving their physical surroundings, but in the process exacerbated existing social inequalities by benefiting some while discriminating against others, particularly low-income, transient, and Native American populations living in the low-lying areas of the city. Even the triumph of solving Lake Washington's pollution problems through the creation of a regional governmental body had uneven social impacts and unintended consequences. However, the rise of Metro would signal a turning point for the city and pave the way for new conceptions of human/nature relations beginning in the 1970s. In the next chapter, I examine how government and nongovernment actors have shaped urban water flows in Seattle in the last four decades, building upon the Promethean foundation created by Thomson and his successors while developing new interpretations of urban nature.

6

Reasserting the Place of Nature in Seattle's Urban Creeks

In March 1999, the U.S. National Marine Fisheries Service announced that eight wild salmon and steelhead populations in the Pacific Northwest would be listed as threatened under the Endangered Species Act, and a ninth population would be listed at the highest level of endangered. The listing was the broadest action in the twenty-six-year history of the Endangered Species Act (ESA) in terms of geography and impact on human populations, due to the inclusion of the metropolitan regions of Portland and Seattle. The 1999 ESA listing reminded Seattleites of the enduring crisis of salmon management in the Pacific Northwest since the late nineteenth century and reinforced the notion that salmon recovery was not merely a rural issue to be solved with increased hatchery activities and preservation of wilderness areas but through restoration of the degraded waterways in Seattle. It was time to bring nature home again.

Almost all of Seattle's waterways, from Lake Washington and the Duwamish River to the smallest creeks and ponds, were significantly modified in the twentieth century to allow for urban development. In the most extreme cases, as with the regrading projects, the waterways were completely filled in or directed into pipes; in others, the banks were regraded and dams and weirs were installed. Today, the creeks of Seattle are a particular form of urban water that provide multiple insights into the relations between the city and nature. These waterways are at once natural and technical, loved and despised, beckoning and threatening, beautiful and ugly, and they receive a great deal of attention and energy from the municipality, environmental and neighborhood organizations, and individual residents.

In this chapter, I examine activities by these groups since the 1960s to reorient Seattle's urban runoff regime and reintroduce nature—specifically, salmon—back into the city. The previous chapter was about the transformation of Seattle from an organic city in the nineteenth century to a rational, planned city in the twentieth century; this chapter focuses on activities to protect and revitalize metropolitan nature and serves as a response to the third phase of the Promethean Project and the crisis of nature at the hands of humans. Some of these restoration activities are part of the local government's commitment to environmental protection; others are grassroots efforts of local residents aimed at reworking urban nature to improve quality of life. As a whole, they entail different avenues of intervention to recast the relations between humans and nonhumans in the city.

Seattle and Urban Environmentalism in the 1970s

The economy in Seattle boomed in the World War II era due to military contracts for Boeing, the local airplane manufacturer; the city would earn the nickname "Jet City" in the postwar decades as Cold War military spending and the new commercial airline business buoyed the local economy.[1] However, the economic boom was short-lived; crisis hit in the late 1960s when the national aerospace economy floundered and Boeing laid off almost two-thirds of its one hundred thousand employees between 1968 and 1971. The unemployment rate in Seattle skyrocketed from 3 percent at the beginning of this period (well below the national average) to 15 percent (double the national average), and the Seattle metropolitan region spiraled into a deep economic recession.[2]

It was during this low point in Seattle's history, the so-called Boeing Bust years, when the laissez-faire business character of Seattle's municipal government would be abandoned. In the municipal election of 1967, voters adopted a new strong mayor form of municipal governance, reflecting the influence of national social movements and antiestablishment politics, while drawing on Seattle's history as a labor-friendly town. Labor strikes and protests were common in Seattle in the 1910s and 1920s; in 1919, the city was the first in the United States to be shut down completely by a strike.[3] This emphasis on "people power" would reemerge with the adoption of the strong mayor form of governance and designation of a leader who could stand up to the corrupting forces of the political machine, where true municipal power was located.[4] The result was a participatory form of urban governance emphasizing citizen advocacy and control of municipal politics with the mayor championing the citizens' desire for self-governance.[5]

The new municipal structure would also serve as a counterpoint to the antitax, anti-Seattle sentiment that emerged in the adjacent suburbs in the 1950s and 1960s over the fight with Metro for regional governance.[6] Ultimate control of urban politics would be directed not by the regional government or by municipally elected officials but by an active citizenry. Related to this turn toward participatory governance, many Seattleites found inspiration in the related environmental movements of back-to-the-land, bioregionalism, and appropriate technology. The emphasis on nature's abundance and the quest for a balance between humans and the landscape resonated with those residents of the Pacific Northwest who revered nature and wanted to live in harmony with their natural surroundings.

Ernest Callenbach's 1975 novel *Ecotopia* was particularly influential in bringing these ideas together into a coherent whole. In Callenbach's fictitious account, Northern California, Oregon, and Washington seceded from the rest of the United States to form a new self-sufficient nation called Ecotopia. Ecotopians rejected the U.S. emphasis on economic growth and progress, reorienting their economy toward biological survival, quality of life, and a balance between humans and nature. Ecotopians championed a mandatory 20-hour workweek and emphasized organic agriculture, sustainable forestry, recycling, and renewable energy strategies. Similar to Appropriate Technology enthusiasts, Ecotopians put technology to the service of humans, rather than the other way around.[7] As Callenbach writes, "Ecotopians claim to have sifted through modern technology and rejected huge tracts of it, because of its ecological harmfulness."[8] Environmentally detrimental products such as the internal combustion engine and plastics were outlawed, and public funding for scientific research was directed toward benign forms of consumption. Ecotopians also practiced a form of ecological democracy that involved participatory governance based on the protection of the environment; all residents were eager to engage in heated but civil debates about political matters and government policies on a daily basis.

For a small but active group of Pacific Northwest residents, Callenbach's novel captured the essence of the region.[9] They already appreciated the promise and potential of the region as a place where nature and humans could live in harmony, and Callenbach provided these residents with a blueprint for transforming their beliefs about environmental protection and participatory governance into reality. A city hall insider and long-time resident of Seattle succinctly notes, "Once Ernest Callenbach did Ecotopia, people realized they could do something different."[10] Reflecting on the synergy between participatory governance and urban environmentalism, historian Jeffrey Sanders writes:

Like Chicago at the turn of the last century, Seattle in the 1970s became a laboratory where diverse actors linked social and environmental change, and where they tested ideas of sustainability on a local scale and in immediate spatial terms. Between 1960 and 1990, Seattleites set about to radically remake the idea and physical form of the city and in the process, they invented a persuasive version of postindustrial Ecotopia, and a model of postmodern urbanism. As a result, as early as the 1980s, Seattle, the largest city in the Northwest, was uniquely poised as the leader of the sustainability movement and a symbol of urban revival.[11]

Seattle's political activities of the 1970s centered around issues of water pollution, freeway expansion, unsightly billboards, and the general empowerment of local residents to make decisions about how they wanted to live and grow as residents of the city. There was an emerging recognition among urban professionals, counterculturalists, and environmentalists that the contemporary city form was more malleable than previously thought.[12] In other words, there were alternatives to the path of economic and societal progress introduced by Thomson and other Progressive reformers in the late nineteenth century.

Local environmental activities were inspired by self-sufficiency, do-it-yourself activism, sustainable agriculture, and organic gardening through groups like Seattle Tilth and the Puget Consumer's Cooperative, as well as the work of local writers and artists.[13] Of particular note was University of Washington architecture professor Victor Steinbrueck, who gained local notoriety for galvanizing Seattleites to preserve historic parts of the city in the early 1970s, including the now-cherished downtown destinations of Pioneer Square and Pike Place Market. Steinbrueck was also influential in reinterpreting the Seattle landscape as a place of both humans and nonhumans. He published two popular books in 1962 and 1973 with sketches of Seattle's cityscapes that included buildings, hills, streets, vegetation, and water, forwarding a vision of a green city where people and nature existed in harmony (see figure 6.1).[14] Similarly, writer Janice Krenmayr published a column in the *Seattle Times* Sunday Magazine called "Foot-loose in Seattle" that focused on experiencing the city through urban hikes. She built upon Steinbrueck's vision of metropolitan nature to include activities that would allow urban residents to experience their hybrid surroundings.[15] Their work, along with a number of other writers and artists, reframed Seattle not as a city *surrounded by* nature but as a city *of* nature, and envisioned a new relationship between humans and nature that would be less focused on control and more on partnership and stewardship.[16]

Figure 6.1
A sketch from Victor Steinbrueck's Seattle Cityscape shows the relationship between the built and natural environments in Seattle. *Source*: Steinbrueck 1962.

Participatory Planning and Growth Management

The participatory planning model that began as a grassroots effort in the 1970s would become the de facto approach to city planning in subsequent decades. The so-called Seattle Way emphasized process, collaboration, and consensus and was fueled by the strong support of a string of mayors, including Charles Royer (1978–1989), Norm Rice (1990–1997), and Paul Schell (1998–2001). These mayors institutionalized neighborhood power and self-actualization by encouraging the creation of "little City Halls" and directed significant amounts of municipal funding to neighborhood groups to undertake their own civic improvement projects.[17] This bottom-up form of urban development is exemplified by the City of Seattle's Neighborhood Matching Fund Program, a successful program that has doled out more than $45 million for over 3,800 community-executed projects since it began in 1988.[18]

In 1990, the state legislature passed the Growth Management Act to address urban sprawl and the loss of rural lands and wilderness areas. The state government was particularly concerned with the expansion of the urban corridor situated to the west of the Cascade Mountains. The City of Seattle responded by initiating a comprehensive planning process, and in 1994 released its twenty-year vision, *Toward a Sustainable Seattle.*[19] The comprehensive plan was based on values of environmental stewardship, social equity, economic opportunity, and community, with an emphasis on channeling development into urban villages and centers. The municipal government then invited neighborhood organizations to develop their own local plans to comply with the goals of the comprehensive plan. Almost 20,000 citizens participated in the development of thirty-eight neighborhood plans over a period of fifteen months in 1998 and 1999 to delineate how each neighborhood would grow over the next twenty years. In short, it was an unprecedented participatory planning effort. Although the neighborhood plans were not legally binding, they required new development proposals that were at odds with plan goals to go through a review and revision process.[20]

Not surprisingly, urban environmental restoration was a significant focus of neighborhood planning efforts. Residents singled out creeks, parks, and open spaces as places where nonhuman elements of the city resided and as important places to restore and protect. In 1992, Mayor Rice acknowledged the significance of nature in the city in his introduction to the two-year Environmental Priorities Program:

We must incorporate an ethic of environmental stewardship into everything we do. Some people think that the words "urban environment" are a contradiction in terms. I disagree. While our city is widely—and justifiably—recognized as a leader in urban environmental management, there are opportunities for improvement and actions we can take that would make Seattle even more of a model for other cities. . . . I consider environmental protection and enhancement to be an integral piece of the overall urban agenda. It is not separate from our efforts to improve our schools, our neighborhoods, our economy, our transportation system, and our public safety—it is *part* of them.[21]

Rice embraced a relational perspective that recognized humans and the built environment as part of rather than separate from nature. He understood the growing importance of urban creeks to those residents who lived adjacent to them, singling out urban runoff as "perhaps the most pervasive and difficult to control water quality problem" in Seattle.[22] Reflecting on this time period, a municipal staff member notes, "There was a tremendous amount of mobilization around particular creeks because they were local, they were literally their backyards."[23] The creeks would become a perfect place to direct the energies of self-governance and reverence for nature.

Urban Runoff and the Creeks of Seattle

The biological and physical condition of the creeks in Seattle is closely related to how the city developed its drainage networks over time. As noted in the previous chapter, the central core of Seattle continues to be served by the nineteenth century combined sewer network designed by Benezette Williams. Combined sewer networks serve about one-third of the geographic area of the city, carrying a combination of sanitary and stormwater volumes to Metro's regional wastewater treatment plants (see figure 6.2).[24] A second type of drainage network is informal and consists of ditches and culverts that cover another third of the city's geography. These networks are prevalent in the northern areas of the city that were annexed primarily in the 1950s. The municipal government agreed to upgrade these facilities as part of the annexation process, but due to funding constraints, significant infrastructure upgrades have been completed only on major arterials. And finally, separated or partially separated networks handle some or all stormwater volumes separately from sanitary wastewater volumes in the remaining third of the municipality's jurisdiction.[25] With the partially separated networks, roof drains are directed to the sanitary sewer and street drains flow to the storm sewer. The heterogeneous runoff regime in Seattle can be understood as an ad hoc evolution of drainage logics applied by the municipal government over time to manage the city's stormwater volumes.

The ditch-and-culvert networks and the separated or partially separated networks discharge directly into the city's creeks. The creeks serve as the trunk lines for these networks because they are low-elevation features that provide an existing and inexpensive conveyance channel for urban runoff. These drainage networks are an ecotechnic hybrid that bind land development to waterflows and their biota, including salmon. Seattle once had some forty creeks that sustained healthy salmon populations; the conversion of these creeks to drainage conduits, along with loss of habitat due to urban development, has resulted in only four creeks that can sustain salmonid populations today.[26]

Like most U.S. cities, the municipal government in Seattle has emphasized conveyance and flood protection when designing, constructing, and operating the three types of drainage networks. Seattle Public Utilities (SPU) was formed in 1997 to combine the municipal government's operations of water supply, wastewater conveyance, stormwater management, and solid waste management into one organization.[27] SPU's drainage activities are divided into four major programs: stormwater and flood

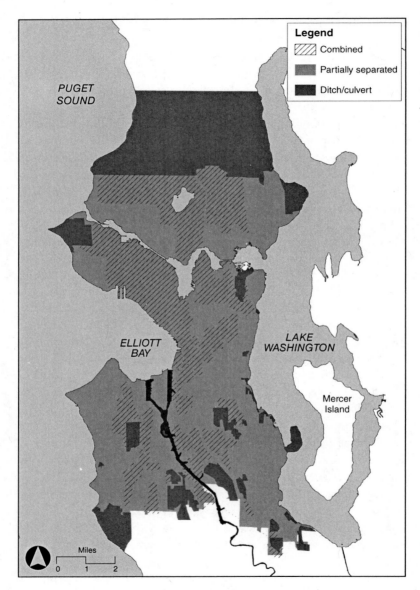

Figure 6.2
A map of the three drainage networks in Seattle. *Source*: Derived from SPU 1995, Map 1-1.

control, landslide mitigation, aquatic resource protection—water quality, and aquatic resource protection—habitat. SPU has issued three Comprehensive Drainage Plans to date (1988, 1995, and 2004). Each of the plans place an increasing emphasis on water quality and habitat protection due to a variety of state and federal laws relating to environmental protection, notably the Clean Water Act and the Endangered Species Act.[28]

It is only since the late 1990s that the municipal government has officially recognized urban creeks as an element of the city's drainage networks and has made a concerted effort to improve water quality conditions and restore habitat. However, there have been many nongovernmental efforts since the 1960s to change the conveyance logic of urban drainage in the city, initiated by community activists who want to improve the conditions of their neighborhoods. The activities by community activists to reorient drainage patterns in Seattle suggest that the municipality is not the only urban environmental manager. The grassroots work of residents reorients urban nature in ways that are strikingly different from the municipality's technomanagerial approach.

Restoring Nature in the City: Carkeek Park

One of the earliest and most noted neighborhood projects focusing on urban creek restoration in Seattle involves Piper's Creek in northwest Seattle. The Piper's Creek watershed is in the Broadview neighborhood and is the third largest in the city, comprising 2.5 square miles of drainage area. The watershed is unique because of its multiple high-gradient tributaries couched in steep ravines, with much of the creek located in Carkeek Park. The remainder of the watershed is dominated by residential land use with an overall impervious surface coverage of 57 percent (typical for Seattle's residential neighborhoods) and relies on a ditch-and-culvert drainage network.[29]

Carkeek Park opened in 1928, eight years after the last old growth timber was logged from the land and a year after the last salmon were spotted in the creek.[30] In subsequent years, the municipality improved the park with a series of trails and shelters, supplemented in the 1930s with help from the federal Civilian Conservation Corps program. Creek restoration activities began in 1965 when neighborhood resident Nancy Malmgren led a Girl Scout group on an expedition in the park and discovered a severely degraded creek. Malmgren initiated an informal volunteer creek restoration program and her persistent efforts eventually led to the formation of the Carkeek Watershed Community Action Project (CWCAP) in 1979.

CWCAP expanded Malmgren's mission by conducting outdoor education activities and hosting special events, and in the early 1980s, initiated a salmon enhancement project with funding from a Metro grant to introduce hatchlings into the creek.[31] In 1986, the Piper's Creek watershed was selected by the Washington Department of Ecology as an early action watershed project to control nonpoint source pollution, the only urban watershed of twelve selected in the Puget Sound region.[32] In 1990, the Piper's Creek Watershed Action Plan was approved by the City of Seattle and the Department of Ecology.[33] As small salmon returned to the creek, the thousands of volunteer hours devoted to restoring habitat as well as over $2 million in government grants from King County and the state government finally began to produce results.[34] Piper's Creek serves as an early example of the potential for urban creeks in Seattle to be rehabilitated and today, Malmgren is a local legend of urban environmental restoration.[35]

Carkeek Park consists of a series of beautiful, lush trails through steep ravines with multiple tributaries (see figure 6.3). Upon entering the park,

Figure 6.3
Piper's Creek in Carkeek Park with logs placed by volunteers to provide salmon habitat (left), the waterway surrounded by lush vegetation (top right), and a sign announcing the return of salmon (bottom right).

the busy and loud urban landscape quickly dissipates and one is surrounded by what appears to be pristine wilderness. Indeed, there is little
evidence that the large trees in the park are secondary growth and the creek
has a small but consistent salmon population that returns every year in
early winter. Recreating prehuman nature in the park has been the goal
of Malmgren and her cohorts, and they have done a remarkable job; the
park today is one of the few places where one can experience "wilderness"
inside the city limits.[36]

The success of creek restoration in Carkeek Park is tempered to some
degree when one recognizes that the restoration efforts do not extend
beyond the park boundaries. As such, the park reinforces the separation
between humans and nonhumans by creating a distinct place where nature
dominates and human visitors are encouraged to "tread lightly" and "look
but don't touch." The restoration efforts were possible only because this
land was unsuitable for residential development; otherwise, it is likely
that it would have been developed like the surrounding neighborhoods.
In many ways, this marginal landscape has become a historical recreation
of romantic nature lost to local residents in the late nineteenth century.
Indeed, nostalgia for a less urban, quieter, and more benign city seems
to run deep in the environmental restoration logic followed here. This is
not to argue that the Piper's Creek restoration efforts are undesirable but
rather that they follow a specific conception of human-nature relations,
one that follows the conventional logic of separating humans from nonhumans to create seemingly purified forms.

The restoration of salmon populations in Piper's Creek is also notable
for how urban nature was reworked. Grassroots action over decades centered on a particular part of the city, one that has been largely neglected by
the municipality. The neighboring residents developed a collective vision of
an improved future and then collaborated with various government entities to fulfill this vision. This action has also occurred in other noteworthy
projects in Seattle, such as Ravenna Creek, Taylor Creek, and Fauntleroy
Creek, where residents have married their love of place and love of nature
into action-oriented activities.

The Municipality's Natural Drainage Systems Approach

The burgeoning support of urban creek restoration among urban residents served as a springboard for institutional reform within the municipal government. When Mayor Schell came into office in early 1998, he
recognized the upcoming millennium and the imminent ESA listing as an

opportunity to establish historic municipal government programs. Schell had a unique background for a mayor; in the 1970s, he served as director of community development for the municipality, and then worked as a real estate developer, a Port of Seattle commissioner, and acting dean of the College of Architecture and Landscape Architecture at the University of Washington. More striking than his eclectic résumé was his outspoken support of urban sustainability. Schell was fond of invoking a combination of Callenbach's *Ecotopia* and bioregionalist idealism when he speculated on the possible future of the Pacific Northwest and was a self-proclaimed visionary who enjoyed philosophizing about big ideas.[37]

Even more than his predecessor, Norm Rice, Schell recognized that there was a connection between neighborhood activities and the environmental character of the city, and he was determined to leave a legacy to reflect this connection, stating, "The citizens who settled Seattle wanted to conquer nature. We want to celebrate it."[38] Clearly, he was a student of Seattle's development history and recognized an opportunity to create an alternative to the Promethean goal of the ultimate control of nature. Anticipating the forthcoming 1999 ESA salmon listing, Schell wrote an op-ed piece in the *Seattle Times* titled "Saving Salmon May Save Ourselves":

The reasons for a declining salmon population can be summed up easily: We humans create too many competing uses for our rivers, streams and oceans. If you're looking for something to blame, it's the growth and development that surrounds us. . . . In short, millions of people have crowded out millions of fish. It's time to strive for a better balance. . . . Our wild salmon connect us to the land we live on, to the people who were here before us, and to the people who will be here when we are gone. We can save this wonderful creature, and by doing so, we may very well be saving ourselves.[39]

Schell's conflation of salmon health and human health was a reflection of his belief in the standard prescription for sustainable development with environmental protection and economic development as complementary rather than competing goals.[40]

Schell put this belief into action with the inauguration of the Urban Creeks Legacy Program in 1999, designating four creeks—Longfellow, Piper's, Taylor, and Thornton—in the four corners of the city to be targeted for restoration efforts (see figure 6.4). The initiative dedicated $15 million in funding over a five-year period to improve urban salmon habitat. A former director of SPU notes, "The idea was to have fish come back to the four corners of the city. We're not going to make a dent in saving salmon, but we're going to make a big dent in the public's understanding of the life of salmon and the life of a creek."[41] The Urban Creeks Legacy

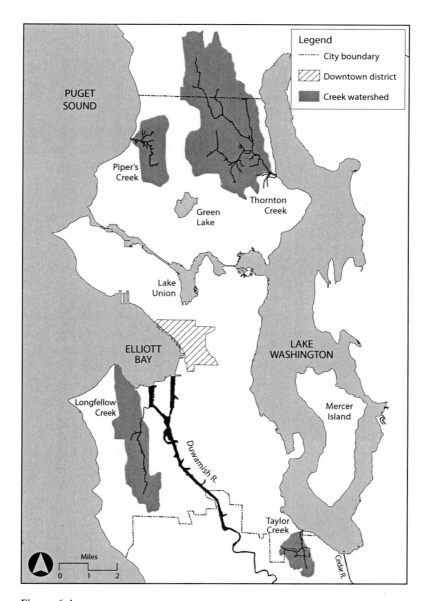

Figure 6.4
The four waterways of the Urban Creeks Legacy Program.

Program would be a municipal approach to reveal the connections between nature and Seattle residents, even if the rehabilitation of urban salmon populations was not a realistic outcome. The approach represented a unique form of environmental protection whose end goal was public education and infrastructure improvement rather than biological restoration and protection. Schell effectively translated the metropolitan nature rhetoric from the late nineteenth century to fit with emerging notions of sustainable urban development while tapping into the growing enthusiasm for urban creek restoration that began in the 1960s. Reflecting on the importance of urban creek restoration activities in Seattle, a local journalist argues: "It turns the environment from an abstract ideal to a stomping ground city people can see every day. Seattle's creeks are potential refuges not just for salmon but for creatures that range from deer and eagles to algae and aquatic insects. They are places of meditation and retreat. They are outdoor classrooms. They are filters for runoff. They are corridors for both hikers and wildlife. They are where we teach our kids to pick a future."[42]

To initiate the Urban Creeks Legacy Program, Schell asked SPU staff members to develop a pilot project that would reduce the impact of urban runoff on receiving creeks.[43] SPU staff recognized that the City of Seattle was almost completely built up and rather than focus on greenfield protection, the most promising opportunity for intervention lay in leveraging the public rights-of-way that comprise 25 percent of the city's land area.[44] It was here, in the public corridors of the city, where SPU could apply its technical expertise with municipal funding to rework the relationship between Seattleites and nature. Rather than continue to rely upon the conventional engineering approach of taming and controlling nature, this was an opportunity to develop a new logic of urban drainage that traced the problems and solutions of urban drainage back to the source. The result was the Natural Drainage Systems (NDS) approach.

The NDS approach mimics the preurban forested condition as closely as possible but does not adopt the aesthetic look of the forest.[45] Similar to Low Impact Development and source control strategies mentioned in chapter 2, the NDS approach involves strategies of infiltration, flow attenuation, filtration, bioremediation with soil and plants, reduction of impervious surface coverage, and provision of pedestrian amenities. It integrates expertise from civil and environmental engineering, landscape architecture, ecological planning, and biology to reimagine the water metabolisms of the city. The emergence of source control as an alternative drainage strategy suggests that urban water flows should not be directed by *hydraulic calculations* and engineered treatment systems but rather close analysis of *hydrologic function* and the leveraging of ecosystem services.[46] An SPU staff member states, "NDS signified a shift from the conventional stormwater approach, which is all about getting water from point A to point B, to saying that SPU does more than just convey water. We do habitat work and water quality work."[47] Reflecting on the impetus for the NDS approach, another SPU staff member notes: "We already had a lot of interest in the creeks and there was an opportunity because it was the upcoming millennium. We convinced the mayor [Paul Schell] that instead of building some fancy building like in London, we should restore our creeks and trees. So we got this huge influx of money in 1999 to do projects and that's how the idea gelled to try the NDS approach. The money was flowing, the economy was good, so we got the opportunity to try this."[48]

The NDS approach is an evolutionary step in the Promethean approach as developed by Thomson and his contemporaries in the first half of the twentieth century. It builds upon the bureaucratic structure and centrality of scientific and technical experts, supplementing their knowledge with the ecological and biological sciences (similar to the Lake Washington cleanup) as well as a significant emphasis on the design disciplines. Though the character of intervention is markedly different, the NDS and Promethean approaches share a faith in the underlying structure of techno-managerial governance.

The first project to follow the NDS approach was SEA Street (SEA is an acronym for Street Edge Alternative), a small retrofit project on a single block in the Piper's Creek watershed directly upstream from Carkeek Park. An environmental activist plainly states that "the high quality of Piper's Creek and the downstream quality of Carkeek Park was a key factor in SPU choosing the location for the SEA Street project."[49] SPU staff members also targeted the Piper's Creek watershed because it had a ditch-and-culvert network for stormwater drainage and they could thus justify the project not only as a retrofit but also an upgrade to a street with antiquated infrastructure.[50] SPU staff identified several neighborhood blocks of appropriate slope and length for the pilot project but also sought strong support from the residents who would be directly affected. They held a contest and the winning block on Second Avenue NW between NW 117th and NW 120th Streets had eighteen out of nineteen residents voicing enthusiasm and support for the project.

SPU staff members oversaw a collaborative design process with the residents (reflecting the Seattle tradition of neighborhood self-governance), and the design team negotiated the various issues such as street and

Figure 6.5
The SEA Street project with a meandering vehicle lane (left), ribbon curbs and a narrow sidewalk (top right), and bioswales (bottom right).

sidewalk width, curb height, and number of parking spaces, as well as runoff pathways and locations of swales. The existing street width was reduced from twenty-five feet to fourteen feet, a significant number of parking spaces were removed, and a series of bioswales along with a four-foot-wide sidewalk were constructed on the west side of the street (see figure 6.5). In some cases, the swales were lined with an impermeable barrier to avoid basement flooding issues. Finally, a landscape architect worked with homeowners to integrate the street edges with their properties and select from a palette of native and nonnative plants.[51]

The project was completed in 2001 and included 660 feet of infrastructure upgrades at a cost of $850,000. Reflecting on the cost, an SPU staff member notes, "It was a small cost compared to other SPU projects but given the fact that it was a prototype, it was heavily scrutinized."[52] The change in the street character was dramatic, transforming a nondescript two-lane residential street with a ditch-and-culvert drainage network into a meandering, narrow road with lush vegetation and a series of swales. The design slowed vehicle traffic, provided a sidewalk for pedestrians, and anecdotal evidence suggests that residents have benefited from higher property values.

In terms of hydrologic performance, the new street design reduced impervious cover by 11 percent. Researchers from the University of Washington's Department of Civil and Environmental Engineering monitored the drainage performance in 2001 and 2002 and noted that "the project has the ability to attenuate all or almost all runoff over a fairly wide range of conditions."[53] Except for very heavy and infrequent rainfalls, the swales are able to infiltrate all stormwater runoff. Furthermore, stormwater flow velocities downstream in Piper's Creek were reduced by 20 percent and the system has withstood a number of significant winter storms in subsequent years.[54]

There was initial concern by some SEA Street residents that the prototype design was inferior to the traditional curb-and-gutter design and thus did not constitute a genuine street improvement; there were also complaints about the reduced number of parking spaces on the street.[55] However, the negative perceptions of the project have largely dissipated as residents on the street (as well as adjacent streets) have begun to use the right-of-way in new ways.[56] An SPU staff member notes, "From a social perspective, the use of the street has changed. It has turned into essentially a public walking space. People with kids and their training wheels perceive it as a safe place to go for a walk. It's dramatic."[57] SEA Street is not merely an upgrade to the city's drainage network; it completely reinvents the aesthetic and functional qualities of public rights-of-way. The project serves the dual purpose of restoring the natural hydrologic function of the urbanized watershed while increasing residents' understanding of the natural processes in which they live.[58] It is a noteworthy example of what landscape architects call "eco-revelatory design," in which the goal is to highlight the connections between the human and nonhuman through a process of revealing and marking.[59]

The reimagined streetscape restores some of the historical functions of urban streets from times before streetcars and automobiles transformed these social spaces into transportation conduits. Reflecting on the importance of public spaces such as streets and sidewalks, environmental philosopher Avner de-Shalit argues that "they represent the city, its culture, its self-image, and its thick idea of the good more than anything else."[60] Urban theorist Stanford Anderson goes further, interpreting streets as relational spaces in the contemporary city: "The intermediate position of streets in the environment, intersecting public and private, individual and society, movement and place, built and unbuilt, architecture and planning, demands that simultaneous attention be given to people, the physical environment, and their numerous interrelations."[61]

More difficult than negotiating with homeowners over the new street design were the battles that erupted between municipal departments over the modification of right-of-way areas. In particular, the transportation and emergency services departments cited concerns about the safety of children with respect to the open drainage swales as well as access for emergency vehicles and adequate provision for parking.[62] One SPU staff member notes, "For SEA Street, the transportation department only approved the width of the street because it was a demonstration. They did have an emergency response there and they've been down the street, but they don't like it."[63] SEA Street posed a direct challenge to existing public right-of-way conceptions but had the advantage of being a project initiated by the municipality; thus, SPU had more negotiating power with other municipal departments and could successfully push for more radical reinterpretations of street function. The battle between municipal experts was won by the stormwater engineers, largely due to the support of the mayor and the promise that this would be a one-off project—an experiment, rather than a new approach to street design.

The SEA Street project has received regional and national acclaim as an innovative approach to sustainable urban stormwater management. In 2003, the project and the NDS Program received the "Vision 2020 Award" from the Puget Sound Regional Council for promoting a livable Pacific Northwest region. And in 2004, the NDS program won the prestigious Innovation in American Government Award from Harvard University's Kennedy School of Government as well as $100,000 in prize money.[64] Riding on this success, SPU would go on to replicate, expand, and revise the SEA Street process with four more projects (see table 6.1).

Table 6.1
Natural Drainage System projects, 2001–2009

Project	Year completed	Watershed	Size (blocks)	Drainage area (acres)
SEA Street	2001	Piper's	1	2.3
110th Cascade Project	2003	Piper's	4	21
Broadview Green Grid	2005	Piper's	15	32
Pinehurst Green Grid	2007	Thornton	12	49
High Point	2010	Longfellow	34	129

Source: Seattle Public Utilities 2008.

It was not a one-off project by any means but a new approach to mediating the tensions between humans and nature in the city through creative engineering intervention. The subsequent projects were selected not only for their potential to upgrade existing infrastructure networks but also their promise for improving the biological integrity of downstream waterways.[65] Furthermore, the NDS approach represented a significant shift in the governance of urban nature with design and public involvement as central components.

The Grid and the Swale: The High Point Redevelopment

Three of the subsequent NDS projects—110th Cascade, Broadview Green Grid, and Pinehurst Green Grid—are similar to SEA Street in their focus on reorienting public rights-of-way in residential neighborhoods, but they expanded the model to include larger areas. Two of the projects are in the Piper's Creek watershed and the third is located just east in the Thornton Creek Watershed; all are in areas with ditch-and-culvert drainage networks and are justified as infrastructure upgrades. The fourth project, the High Point Redevelopment, is much more ambitious in scale and represents a significant departure from the other projects.

The High Point Redevelopment project is a complete redevelopment of a 129-acre site in the Longfellow Creek watershed of West Seattle. High Point was originally built in 1942 as temporary housing for workers in the burgeoning local war economy and in 1952 was converted into a 716-unit public housing development owned and managed by the Seattle Housing Authority (SHA). In the early 2000s, SHA decided to redevelop the community and initiated a master planning process to create a walkable, multiuse, New Urbanist neighborhood with modest green building strategies.[66] As luck would have it, SPU staff members were just coming off their initial success with SEA Street and were looking for an opportunity to scale up their approach to a larger project. They saw the High Point Redevelopment, encompassing almost 10 percent of the Longfellow Creek watershed, as an ideal opportunity and approached SHA about a possible collaboration.

SHA agreed to incorporate the NDS approach at High Point under three conditions. First, SPU would pay the difference between the conventional drainage network cost and the upgraded NDS network. SPU was willing to pay for this performance upgrade because it would partially fulfill the goals of the Urban Creek Legacy Program for Longfellow Creek and would serve as another example of the efficacy of the NDS approach. Second, SHA required that SPU be responsible for obtaining municipal approval for the various building permits for the project, a formidable effort. This point was also agreeable to the SPU staff because they were an inside municipal government agency, unlike SHA, which is a public corporation affiliated with but separate from the other municipal departments.[67] Third (and perhaps most importantly), SHA wanted to ensure that the appearance of the project would not be affected by the NDS approach, following on the aesthetic philosophy of "normalcy" as promoted by New Urbanist designers.[68] An SHA staff member notes, "The key point to me was that it had to look like a regular neighborhood. I really felt strongly about that. I knew about these Bill McDonough designed communities where you couldn't tell the front door from the backdoor. It was really important to make it look like a regular place so we wouldn't be experimenting with low-income residents."[69]

SPU agreed to these conditions, and over the next three years, the design team worked hand in hand to reconcile the various goals of the project to create a high density, walkable, environmentally friendly, and affordable community. The level of collaboration between the various city agencies, particularly in designing the thirty-four blocks of right-of-way, was unprecedented.[70] A staff member in the City of Seattle's Department of Planning and Development notes:

There were endless hours of regulatory meetings within the City of Seattle because we changed every standard. You're not even allowed to have roof downspout disconnects in the city. With new development you're supposed to tie in to the storm drain. Just to have the stormwater go to a splashpad and surface drainage was a big change. Having the curb go from six inches to four inches involved hours of meetings. Every little thing was a variation from the standard. Without incredible commitment from my colleagues at SPU and support from upper management, it would have fallen through easily. SHA also put in tons of effort but we were charged with putting all of the changes through all of the regulations because we were the ones asking for something different.[71]

The extended negotiations over changing the drainage regime at High Point reveal the dominance of conventional stormwater logic, not only in the practices of design professionals and municipal experts but also in municipal development code. The design team was fortunate to include not only dedicated municipal staff members who were willing to challenge the existing standards but also the leading sustainable architecture and engineering firms in the city.[72]

In some cases, the project goals of creating a dense urban environment of New Urbanist development were in direct contradiction with the

NDS goals, reflecting the struggle between ecological and neotraditional designers over the last three decades.[73] For instance, street widths were reduced to twenty-five feet from the standard thirty-two-foot width to create a more walkable and historic feel to the streetscape. This change created more intensive development and reduced the amount of space in the right-of-way for drainage strategies. The design team had to scrutinize the placement of each driveway, parking space, fire hydrant, and curb cut to balance the competing demands in the minimized right-of way area. In addition, the inclusion of the NDS approach to accommodate rather than convey stormwater runoff would have spillover impacts on all facets of the neighborhood's design. At times, there was an undercurrent of resentment by the design team that the project was being compromised by the NDS approach, but eventually the team members found amenable design solutions to satisfy the numerous and sometimes conflicting project goals.[74]

Housing density at High Point is about sixteen units per acre (about double the surrounding neighborhoods) and includes 1,600 apartments units and houses to serve 4,000 residents in a mix of low-income and market-rate rental units as well as market-rate for-sale units.[75] The overall impervious surface coverage of the project is 65 percent, which is high for a residential neighborhood and significantly higher than the other NDS projects.[76] The first phase of the project was completed from 2003 to 2006, and the second phase was completed in 2010; even before construction began, the project received numerous local, regional, and national awards for its innovative design, the most prestigious of which was the Urban Land Institute's 2007 Global Award for Excellence. Today, the project is internationally known as one of the most successful attempts to date to combine social and environmental sustainability goals in an urban context.[77]

With respect to drainage, the design reflects a hybrid of the NDS approach and a conventional curb-and-gutter network. Narrow streets slant to one side rather than from the center to each shoulder, and curb and gutter is used to direct stormwater to twenty-two-thousand linear feet of bioswales, where it is absorbed and filtered. Some of the swales were designed to be shallower than the standard NDS swale to provide more play area for children.[78] SPU developed a *Technical Requirements Manual* that spelled out the philosophy and design implications for the site developers. The final design includes compromises for all involved actors, but overall, the team felt that it achieved all of its goals. Despite the documented performance of the previous NDS projects, the state Department of Ecology questioned the efficacy of the NDS approach for larger storms and ultimately required the design team to include a traditional catch basin and conveyance network as well as a large detention pond as a backup stormwater network. Project team members predict that the conventional network will likely turn out to be unnecessary and that the large detention pond area will eventually be redeveloped as a park or additional housing. The overall increase in cost for the drainage upgrades was $3 million, and the new network is expected to reduce runoff from a one-inch storm by 80 percent when compared with a conventional network.[79] In other words, the majority of rainfall from a typical storm will be infiltrated rather than discharged to Longfellow Creek.

The master planning process included design of not only the rights-of-way but also the private properties in the development, giving the design team the opportunity to extend the NDS approach beyond publicly owned land. Some of the single-family houses have rain chains that provide a visible flow of water as it goes from the roof to the lawn where it is infiltrated, a form of technological transparency or ecorevelatory design (see figure 6.6). Other residences have downspouts that are connected to buried

Figure 6.6
Drainage details at the High Point Redevelopment include curb cuts and bioswales (top left and right) and signage describing NDS strategies (bottom left).

perforated pipes to gradually infiltrate rainwater into the subsurface. However, it remains to be seen how the infiltration features on private property will be maintained over time. The SPU team developed an informational packet for homeowners outlining the drainage features and is depending on the homeowners' association to serve as an educator of new residents in the future. Furthermore, the project team included drainage restrictions in the subdivision recorded documents and future homeowner association covenants to address impervious cover and landscaping requirements.[80]

A significant advantage of the High Point project is that there were no community members; existing low-income residents were moved to other SHA facilities. An SPU staff member notes, "With our previous NDS work, every project had at least one person on each block that thought our approach was the dumbest idea in the world. With High Point, it was different because it was all new; there were no existing homeowners."[81] SHA held a number of meetings with residents from adjacent neighborhoods to address traffic, density, and aesthetic qualities of the redevelopment, but these residents were not central stakeholders in the project. However, the novelty of the design required additional meetings and training for the housing developers. A City of Seattle Department of Planning and Development staff member notes: "In applying LID practices, we had to go through up to three review cycles with developers who didn't know how to do these new approaches from an engineering standpoint. And it was contentious after the first review cycle to tell them, 'Hey, you still didn't get it right—here's what you have to do.'"[82] In other words, there was a steep learning curve for the Phase I construction team. Reflecting on this first team of contractors, another City of Seattle Department of Planning and Development staff member notes: "The contractors are really good at building houses, but I don't think they've ever understood how the swales really function and how important it is to protect them. A great example is that when they were done with a street, they cleaned it up by dumping extra soil into the new bioswales because that's what you do with extra dirt. But the soil will clog up these systems. And they didn't understand how to protect the curb cuts from construction runoff. They just sort of fought it the whole time."[83]

A larger aim of the intensive design process at High Point was to address and correct the restrictions in the existing municipal code and make changes so the NDS approach would be allowed (but not required) for future projects. What was once a pilot project approach on SEA Street was gaining momentum as the standard approach to manage urban runoff in Seattle. The large size of High Point created an opportunity to rewrite the codes that govern urban drainage in the city and permanently change the relations between water and the city's residents. The extended negotiations over urban runoff at High Point reveal the pervasiveness of conventional environmental management, not only in the practice of design professionals and municipal experts but also in municipal regulations. The Promethean logic of controlling nature through big engineering was ingrained in the DNA of the city through its land use development codes.[84] An SPU staff member states, "Now, we're working on how to change the regulations on the books so other developers can do this more easily. So if people are willing to do it, it is allowed. We don't want it to take that much time and effort again because we can't expect anyone to go through that. It's too expensive to go through all of those meetings."[85]

The emphasis on changing the development codes demonstrates how the NDS approach continues to rely on the existing formal mode of technomanagerial governance. However, this emphasis on code revision rather than experimentation at High Point sacrificed many of the social learning aspects of the NDS approach.[86] Specifically, High Point has the look and feel of a traditional residential development; the ecorevelatory character of the other NDS projects is less apparent here. An SPU staff member says, "When we did Broadview and SEA Street, it was definitely shock value and people were accepting of it; leery but accepting. At High Point, they didn't want it to look different, they wanted the signs of affluence which includes lawns that are mowed and green, and straight lines and curbs. It's trying to fit in resident perceptions with what we're trying to do from an environmental standpoint."[87] As such, the NDS approach loses some of its educational benefits in favor of a "have your cake and eat it too" approach of simultaneously achieving social and environmental performance. It was an explicit requirement of SHA to create a neighborhood that had a "normal" feel rather than experimenting with low-income residents, suggesting that ingrained cultural preferences for a finished streetscape with curb-and-gutter design may be even more difficult to overcome than conventional engineering practices and the complex network of development codes. When compared with the other community and municipal urban runoff reform projects in the city, the project tends to be less about revealing the interrelations of humans and nonhumans in favor of a surface aesthetic overlaid on an ecologically functional subsurface, reinforcing the modernist dichotomy of form and function. The high-performance infrastructure is subservient to the social aspirations of the project; the aesthetics of the traditional neighborhood supersede that of water flows.

Furthermore, the NDS approach continues the evolution of techno-managerial governance of urban nature that was established by Thomson and his colleagues in the late nineteenth century and evolved as the municipality formed into a bureaucratic structure dominated by technical experts, regulations and codes, and enforcement activities. The work on SEA Street, High Point, and the other NDS projects demonstrates that this approach can be supplemented with aims of sustainability by opening up engineering practices to more nuanced forms of design that focus on hydrology, ecological services, and even aesthetics. However, there is still an explicit understanding that the expert knows best when reworking the relationship between urban residents and their material surroundings.[88] The political negotiations of the NDS approach were for the most part internal, involving conflicts between the various municipal agencies over different conceptions of the public good. Reworking of urban nature occurred as the experts identified mutually beneficial ways of reinterpreting their bureaucratic missions.

Controversy at Northgate Mall

Neighborhood restoration work and the NDS approach represent two well-known success stories of urban runoff in Seattle. Their motivations for change come from very different origins, but both alter the conventional urban runoff approaches that emphasize conveyance and flood protection. However, not all collaborations between the municipality and Seattleites are as positive as the examples described earlier. The controversy over the redevelopment of Northgate Mall and neighborhood actions to daylight Thornton Creek demonstrate how reworking urban runoff can lead to contentious and highly public battles over the future vision of the city.

Northgate Mall opened in 1950 just outside the Seattle city limits in the heart of the then-burgeoning North End.[89] With eighty stores on sixty-six acres, it was one of the first regional shopping centers in the United States and a daring experiment in retail development at the time. The mall developers wanted to create a one-stop shopping experience to transplant retail from downtown to the suburbs with the lure of free parking. The mall and surrounding area were annexed by the City of Seattle in 1952 and Interstate 5 was completed in 1965, completing the auto-centric retail model as envisioned by the mall owners. Construction of the expansive south parking of the mall involved the burial of about twelve hundred feet of Thornton Creek, the largest in the Urban Creeks Legacy Program.

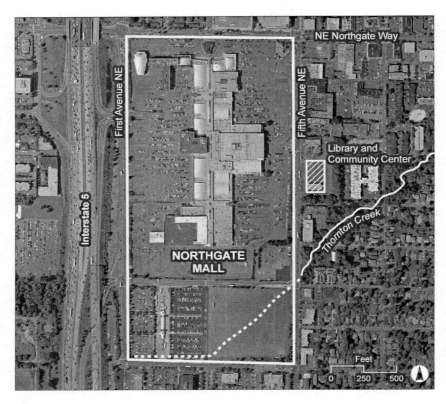

Figure 6.7
Aerial photo of Northgate Mall, showing key features of the redevelopment controversy (the dotted line represents the approximate location of the buried pipe conveying Thornton Creek from Interstate 5 to east of Fifth Avenue NE).

There is some debate about the premall condition of this section of the creek, but most agree that it was not a free flowing channel but rather a boggy wetland that conveyed the creek from west of Interstate 5 to the east of Fifth Avenue NE and eventually to its outlet at Lake Washington (see figure 6.7).

Beginning in the late 1960s, homeowners downstream of the mall began to complain about the deteriorating conditions of the creek and associated flooding problems. The Thornton Creek Alliance was formed in 1992 and—similar to other neighborhood creek organizations—has been active in monitoring, restoring, and educating area residents about the creek. Despite their efforts, Thornton Creek today has a reputation as the one of the most abused creeks in the city due to the intensive development of the watershed. The Thornton Creek watershed is the largest in Seattle at 11.8 square miles and the creek has numerous tributaries that flow past some seven hundred backyards and more than fifteen parks and natural areas, with a total impervious surface coverage of 59 percent.[90]

The controversy at Northgate Mall's southern parking lot began in the late 1980s, when then–mall owner DeBartolo Corporation proposed to double the size of the mall by adding a million square feet of new retail space. Neighborhood groups were adamantly opposed to the expansion, citing a glut of automobile traffic that would clog local streets, and gathered more than a thousand signatures to petition the city council to block the proposed expansion. The council passed a moratorium on development until the Northgate General Development Plan was completed 1993. The plan identified the area as an urban center where development was expected, but specified that future changes should involve the creation of a walkable urban center, a departure from the existing automobile-dominated character of the mall and its satellite businesses.[91] Seattle voters reinforced this vision in 1999 and 2000 by approving over $13 million in bonds for a community center, library, and park on Fifth Avenue on the eastern border of the mall.

Mall redevelopment plans were put on hold for several years until 1998, when the new mall owner, Simon Properties, again proposed a plan to double the size of the mall, with 1.1 million square feet of mixed-use development, including a hotel, a thirty-screen cinema, and additional retail stores.[92] The redevelopment plans targeted the south parking lot, a space occasionally used for RV shows, tent sales, and overflow parking, as the primary site for new development. The city council approved a development plan, and neighborhood residents again protested that the redevelopment would increase traffic in the area. They also argued that the new development would conflict with the goals of the 1993 Northgate General Development Plan as well as forego any opportunity to restore the buried portion of Thornton Creek that ran beneath the south parking lot. They cited the impending ESA listing of Pacific Northwest salmon populations as justification for daylighting the creek.[93]

To resolve the dispute, then-Mayor Schell and the city council hired a mediator in 1999 to find common ground between the battling stakeholders: the mall owner, the municipal government, and environmental and community activists. After hearing arguments from all parties, the mediator determined that the parties were so entrenched in their positions that a compromise solution was impossible. Meanwhile, the fight over the development plans continued, with a new arm of the Thornton Creek

Alliance, the Thornton Creek Legal Defense Fund, and a neighborhood group called Citizens for a Livable Northgate adopting a litigation strategy based on the new ESA listings of salmon populations as justification for daylighting Thornton Creek.[94] The neighborhood coalition took the City of Seattle to court over its approval of the mall redevelopment plans, and after several appeals, reached the state Supreme Court, which declined to hear the case. The mall owner and redevelopment proponents had won the litigation battle, but in the intervening years, the economy had gone into recession; now, Simon Properties wanted to submit a revised development plan that would again require approval by the city council and yet another round of negotiations with the neighborhood groups.

For their part, the neighborhood and environmental activists were not opposed to growth and mall redevelopment but wanted any changes to reflect the neighborhood development plan's primary goal of creating a walkable, vibrant urban center. Furthermore, they saw the daylit creek as a central amenity to a new urban center. One creek activist stated, "We are not sure why there is such obstinance [sic] about looking at designs that will protect the creek. We strongly believe this is a win-win solution, that we can integrate this development with a restored Thornton Creek."[95] They were advocating for Schell's philosophy of sustainable development, which considered economic growth and environmental protection to be complementary goals, and they were willing to stand up to the municipality using litigation to realize their preferred future vision for the city.

Despite the success of the NDS approach and other millennium projects, Mayor Schell would go on to lose his 2001 reelection bid in the aftermath of the notorious 1999 World Trade Organization meeting in Seattle. Greg Nickels swept into office and quickly made a number of significant changes to the municipal planning and governance approach championed by his predecessors. One of his earliest and most contentious decisions was to fire Jim Diers, the popular head of the Neighborhoods Department and a key player for fourteen years in fostering neighborhood governance and power.[96] Nickels attributed the decision to a desire to diversify the leadership in the Neighborhoods Department, but many interpreted the move as a way to shift local political power away from the neighborhoods and back to city hall. A neighborhood activist bluntly states, "Nickels hates neighborhood groups."[97] Nickels's governing approach is often described as autocratic and in the mold of his hometown mayor, Chicago's Richard M. Daley. Nickels exerted his power to centralize environmental governance within the mayor's office, replacing the neighborhood approach of his predecessors with an authoritarian form of municipal politics. He demonstrated that the participatory style of governance as nurtured by his predecessors was fragile and vulnerable; he quickly shifted governance back to the pre-1970s mode of top-down management.[98] Clearly, the Seattle Way of participatory planning was not the Nickels Way.

Despite his dislike of participatory governance and community empowerment, Nickels shared with his predecessors a strong emphasis on sustainability issues and quickly developed an international reputation as a "green mayor" for founding the U.S. Conference of Mayors Climate Protection Agreement. He was also a strong supporter of a variety of municipal sustainability programs focused on green building, urban forests, creek restoration, parks development, and improved mass transit. And like many astute politicians, he developed a reputation for rebranding existing programs and taking credit for them as his own. A city hall insider notes that this sometimes caused problems, stating, "Because Nickels wants ownership, he makes it very hard to do anything that adds value. It limits the ways that we can effectively address these issues because they are attached to the mayor."[99] In the Nickels administration, the Urban Creeks Legacy Program was transformed into the Restore Our Waters Strategy, and his new top-down approach to municipal governance would raise the ire of neighborhood and environmental activists.

Not surprisingly, the Northgate redevelopment controversy escalated when Nickels became mayor in 2001. Nickels had campaigned as a champion of neighborhoods; one of his campaign promises was to daylight Thornton Creek at Northgate Mall. However, he changed his mind upon reaching office, recognizing the property and sales tax benefits to the municipality from a redeveloped Northgate Mall, and in May 2002, he announced his intention to modify the Northgate General Development Plan. A city hall insider states, "Nickels made a lot of promises to the environmental community in his campaign and this was one of the first major breaches of that trust. Neighborhood environmental activists really felt sold out."[100] In response to Nickels's proposal, creek and neighborhood activists formed a new organization, Yes for Seattle, and gathered signatures for a ballot initiative to require developers to daylight urban creeks and restore salmon habitat on major development projects throughout the city.[101] The immediate goal of the initiative was to require Simon Properties to daylight Thornton Creek if and when they redeveloped Northgate Mall, but the long-term goal was more ambitious: to daylight *all* of Seattle's urban creeks. Daylighting assumes that by revealing nature, urban residents will work to protect and cherish their local waterways.

Nickels chafed at the proposed creek daylighting initiative, arguing that it would cost too much money—anywhere from $504 million to $21.6 billion, according to one study by the municipality—and potentially hamper ongoing creek restoration efforts.[102] The conflict pitted economic development versus environmental protection, mirroring the battles over old growth logging that occurred in the Pacific Northwest in the 1980s.[103] The initiative was eventually thrown out by a judge who sided with the city attorney and added that daylighting Seattle's creeks would have widespread negative impacts on economic development in Seattle while doing little to improve salmon habitat in the urban creeks.[104] In March 2003, Nickels announced that he had entered into a development agreement with Simon Properties that would supersede the existing land use regulations at Northgate and pave the way for new development. In the agreement, expansion would occur on the mall's west side, further orienting the mall toward automobile traffic from Interstate 5. In return, Simon Properties would donate 2.7 acres of land on the south parking lot to the city to build a conventional stormwater detention pond over the buried creek. Neighborhood and environmental activists were up in arms, arguing that the mayor was colluding with the mall owner in a closed-door deal and the new agreement went against his promise to daylight the creek as well as the goals of the neighborhood plan to create a walkable urban center.

The mayor's opponents were in luck; his new agreement with Simon Properties required the approval of the city council, the majority of which saw the agreement as a giveaway to economic interests at the expense of local residents.[105] Eight months later, the city council countered the mayor's redevelopment plan with its own plan for redeveloping the site, calling for increased open space and additional community input through the formation of a citizen advisory board. The council's plan was supported by the neighborhood groups and the Thornton Creek Legal Defense Fund, but it was immediately criticized by the mayor, Simon Properties, and the Greater Seattle Chamber of Commerce. The mayor stated that the council's plan would just lead to "more process" and threatened to wait out the terms of two council members that were set to expire in January 2004.[106]

The mayor finally capitulated in December 2003 after the potential property developer, Lorig & Associates, withdrew from the redevelopment project under the threat of even more delays. The mayor agreed to a compromise plan with the city council that would involve the creation of a stakeholder group to shape the redevelopment design and also left open the possibility that the City of Seattle might purchase a 2.7-acre parcel for an NDS type of stormwater facility. The compromise did not guarantee daylighting of the creek but put the option on the table, and Lorig was reinstated as the property developer. In typical fashion, Nickels took credit for the compromise solution in a widely publicized news conference, stating, "The logjam is finally cleared. With today's council actions, we are about to begin a vibrant rebirth at Northgate."[107] The stakeholder group was tasked with identifying common ground between the economic development goals of the mayor and the mall owner, and the community-oriented goals of the residents and city council. Meanwhile, the buried creek would serve as a testbed for integrating environmental restoration and economic development.

In early 2004, SPU was tasked with assessing three different development plants for the south parking lot: a full daylighting scheme that would remove the existing pipe and create an open channel, an NDS approach that would handle surface flows and leave the existing underground pipe in place, and a hybrid approach that would include an open channel while retaining the existing underground pipe. SPU examined the water quality benefits and costs of each option and determined that the hybrid plan would provide the lowest cost and the highest water quality benefits.[108] The Thornton Creek Legal Defense Fund and Lorig & Associates paid for the development of the hybrid daylighting design, relying on the design expertise of a local landscape architect, Peggy Gaynor, to negotiate the criteria of the various stakeholder groups. In June 2004, the hybrid plan received unanimous approval from the stakeholder group and the mayor, with the municipality achieving its water quality and flood protection goals, the community receiving its daylit creek and walkable development, and the developer benefitting from a public amenity for the new development. Most importantly, the development could now move forward after a stalemate lasting almost two decades.

The library, community center, and park were opened in July 2006 to much fanfare, jumpstarting a renaissance of the Fifth Avenue corridor on the east side of the mall. In 2010, Lorig & Associates completed construction of Thornton Place, a five acre mixed-use complex which includes a fourteen-screen movie theater, nearly four hundred apartments and condominiums, and over fifty thousand square feet of retail space.[109] The Northgate Channel became fully operational in September 2009 and includes trails, bridges, interpretive signage, and a vegetated concrete channel located forty feet below the existing grade (see figure 6.8). The majority of water flows through the underground pipe, but a small amount is routed through the channel to provide a constant flow of water.

Figure 6.8
Details of the Northgate Channel, including interpretive signage (left), the waterway situated between the new buildings (top right), and plantings in the concrete channel. *Source:* Cory Crocker.

Much of the success in creating the project has been credited to Lorig & Associates, the developer who agreed to develop the 5.9-acre site when others would not. The firm has a significant amount of experience in public-private partnerships and negotiating among diverse stakeholders to find amenable redevelopment solutions.[110] A local landscape architect notes, "Lorig was the only developer that gave the creek a chance and saw it as an opportunity rather than a barrier to development."[111] The final agreement also has much to do with Peggy Gaynor, the landscape architect who developed the hybrid plan with input from the stakeholder group.[112] A city hall insider notes:

Peggy provided a sense of neutrality; she brought an activist as well as a designer/technical role. . . . She brought an ability to find ways to compromise. When we started, it was all or nothing, an actual daylit creek or an underground pipe. . . . Peggy was able to bring in her expertise to align the community goals with the realities of the project. She brokered a lot of that. Without her, it would have been harder because activists have the vision but they don't have the knowledge to know when to compromise.[113]

Gaynor can be understood as a citizen expert: one who leverages her professional skills as a landscape architect as well as her interest in community empowerment to broker a solution among the competing parties.[114] She was able to navigate the complexities of commercial redevelopment as well the desires of the citizens to have a natural amenity in their neighborhood. Although some stakeholders felt that Gaynor advocated too strongly for the neighborhood activists' insistence on daylighting as opposed to other drainage options, most agree that her hybrid design was the primary reason for the successful compromise. As such, Gaynor used the design process to open up environmental management to multiple competing voices by championing the opinions of local experts, those residents with intimate knowledge of the problems and potential for the creek that are sometimes overlooked by technomanagerial actors. At the same time, she understood the economic and technical aspects of integrating water flows in the built environment and could interpret this knowledge for all parties.

Despite the success in finding a compromise solution, many question the wisdom of spending SPU ratepayer money for a project that has marginal water quality benefits. Another option would have been to pay for the project with money from the city's general fund, but this would have been much more difficult, given the other demands for that funding source. Since 2002, SPU has adopted an approach called Asset Management on all of its water quality projects to account for the social, economic, and environmental benefits and costs of its projects that is commonly referred to as the Triple Bottom Line in sustainable business circles.[115] With respect to the original cost-benefit analysis for the Northgate Channel, an SPU staff member says, "It was close to being a wash. It wasn't just so obvious that we should do this project, unlike others where the benefits are so clear. So it came out ahead but just by a hair, and that was at $10 million. When we go back to do the analysis again, it may be that this time the numbers show it to be a no-go, but we're already committed."[116]

The cost of the Northgate Channel was initially estimated at $7.2 million, the SPU cost assessment was $10 million, and the final cost estimate completed in October 2009 was $14.7 million—more than double the original estimate.[117] Another SPU staff member has a more jaded perspective on the compromised solution:

The whole project isn't about stormwater, it was a political solution to get a development to happen. What's unfortunate is that for the cost of this project, $13 million, you know what you could have bought? That little stretch of channel is not going to help salmon get into that creek. They [neighborhood activists] got so focused on daylighting the creek and getting the money out of the city to do it. They just wanted to win and they did. But so did the mayor. The loser was the ratepayers. They're paying for a project with marginal value that will benefit the new development.[118]

Despite the environmental aims of the neighborhood activists, it is not clear that the Northgate Channel (or any of the urban creek restoration activities, for that matter) will have an appreciable effect on the salmon populations in Seattle. Salmon are frequently the measuring stick for success in these projects, but one SPU staff member predicts that the city's urban runoff projects will be lost in the "statistical noise of water quality and hydrology."[119] However, as stated earlier, the intent of the Urban Creeks Legacy Program and creek restoration more generally is not ecological restoration but public education and awareness. The biological and ecological goals of the Northgate Channel are secondary to larger, more elusive aims of creating new types of integrated social and environmental flows that cannot be measured with traditional economic or biological metrics. The emphasis on education and awareness does not negate the importance of these activities, nor does it suggest that the salmon is a false indicator of environmental health. Rather, this goal incorporates the ideology of salmon as a means to promote urban environmental activities and, in effect, to raise consciousness of residents about their hybrid surroundings.

A local consulting engineer notes, "Daylighting Northgate is totally ridiculous from a fish point of view or a water quality point of view. But if it creates an access that makes the network more visible to people because they see that creek and affects how they view the rest of the system, then it is going to be worth a lot of money."[120] As such, the Northgate Channel is not so much about environmental restoration and salmon protection as it is about fostering an ecological imagination and an environmental ethic among Seattleites. A city hall insider puts the project into perspective with respect to other urban creek projects: "This is a once in a lifetime opportunity. Once that piece of property is redeveloped, it's never going to happen again unless there is an earthquake and everything falls down and the creek decides for itself to come back. For some of these other projects, that doesn't exist right now. You can do them in five or ten years, they're good projects but they're not going away. This one was going away and you have a huge economic impact of redevelopment up there."[121] The Northgate Channel serves an educational purpose by exposing water flows to urban residents. But in order to accomplish this goal, the Northgate Channel has been turned into a highly engineered structure—something very different from the original waterway that flowed through the area and also a significant divergence from the NDS approach practiced by SPU. Daylighting the creek at Northgate meant that the municipality had to revive the Promethean approach of controlling nature rather than taking a more nuanced approach of integrating the waterway into the urban fabric. Make no mistake, the project is referred to as a "channel" and is not by anyone's imagination a creek or even a significantly altered natural waterway. It is a treatment facility that succeeded in unearthing Thornton Creek from its asphalt tomb but is informed by hydraulic calculations rather than the hydrology of the creek.[122]

More important is that the Northgate Controversy highlights the centrality of municipal governance in creek restoration issues and constitutes a significant divergence from the other projects with regard to the central role residents played in the shaping of their communities. Unlike the High Point Redevelopment, in which all stakeholders interpret the final result as a resounding success, the Northgate Channel is tainted by ill will and nasty political infighting rather than a positive outcome. The project is frequently publicized as a highly successful compromise between business and community interests, but the actors involved in the negotiations reflect on the project with disdain due to the overall lack of trust and mutual understanding.[123] This difference is due in large part to all stakeholders beginning the process with entrenched positions regarding the role of nature in cities. The mayor and development interests saw Thornton Creek as an impediment to urban growth and the realization of the full potential of economic opportunity at an inner-city site ripe for redevelopment. Neighborhood activists saw the property as an opportunity to atone for the sins of past development by restoring nature while improving quality of life. Negotiating these competing interests involved back-and-forth accusations and power moves, and it was only in the end that a constructive approach was found to pacify but not entirely satisfy all stakeholders.

Conclusions

The urban runoff and creek restoration activities in Seattle present a wide scope of motivations and approaches that involve aesthetics, community activism, green governance, education, and urban development. As Thrush writes, "Part of Seattle's 'green' persona is a profound ambivalence about its own urban past, perhaps best symbolized by the popularity of community-based organizations and government programs aimed at urban ecological restoration."[124] Many urban runoff activities emerged from a culture of neighborhood governance and urban environmentalism in the

1970s, with a particular emphasis on rehabilitating Seattle's creeks. The neighborhood and municipal actors recast the role of water in different ways, sometimes focusing on the creation of habitat for salmon and at other times working to enlighten urban residents regarding the natural flows that are part and parcel of cities.

A local landscape architect refers to these activities collectively as "unbuilding projects" because they are an attempt to rework the role of nature in the contemporary urban landscape.[125] The ideology of salmon is a means to promote urban environmental activities and, in effect, to raise the consciousness of residents about their hybrid Metronatural surroundings. A local creek activist states, "Without urban salmon, we will have no connection to fish in rural and wilder streams. Fish will become an out-of-sight, out-of-mind situation. It is much better that 10,000 people see 10 salmon than 10 people see 10,000."[126] Getting those ten salmon to return to Seattle's urban creeks can involve important activities that relate humans to other humans as well as their surroundings.

7

Politics of Urban Runoff: Building Relations through Active Citizenship

The urban ecology of the contemporary city remains in a state of flux and awaits a new kind of environmental politics that can respond to the co-evolutionary dynamics of social and bio-physical systems without resort[ing] to the reactionary discourses of the past. By moving away from the idea of the city as the antithesis of an imagined bucolic ideal we can begin to explore the production of urban space as a synthesis between nature and culture in which long-standing ideological antinomies lose their analytical utility and political resonance.
—Matthew Gandy[1]

Salmon and salamanders, conservation land strategies and right-of-way upgrades, passionate neighborhood activists and creative municipal engineers, engineered stormwater ponds and spring-fed community pools: the stories from the previous four chapters reveal the multifaceted character of urban runoff and how water flows connect to a multitude of vexing and intransigent issues facing cities today—population growth and urban expansion, place making and local distinctiveness, social inequity and the right to the city, governance and regulation, environmental degradation and restoration, technological obduracy and change, and so on. It suggests that there are alternative routes or competing pathways to reworking nature in cities and, more broadly, to realize more sustainable urban futures.[2] Urban runoff exposes the contemporary city as a messy agglomeration of overlaps and conflicts, linkages and ruptures. So what can we make of these relations? Can we derive patterns and prescribe generalized courses of action? Or are urban runoff flows specific to place and contingent upon particular social, cultural, and material conditions? Should we champion particular activities over others because of their potential to produce more desirable future conditions, and, if so, by what criteria?

In this chapter, I propose a political framework to describe the urban runoff activities in Austin and Seattle. The notion of "politics" is traditionally used to describe a whole host of activities related to the governance of society—elections, partisan bickering, policy-making activities, fund-raising, campaigning, and so on. I interpret these activities broadly as processes of relation building between different humans actors to find common ground (in the best of circumstances) or to engage in power plays (in the worst cases) and to orient societies into particular configurations. Following on the relational perspective forwarded in chapter 2, I argue that nonhuman actors—water flows, endangered species, ecological services, impervious surfaces, and so on—play a significant role in political activity. They are not merely objects that are manipulated and shuffled around by human actors, nor are they simply the substrate upon which human activities occur; instead, they are part and parcel of the practice of politics. This expansive definition of politics follows upon Latour, who argues that, "In practice, politicians have never dealt with humans, but always with associations of humans and nonhumans, cities and landscapes, productions and diversions, things and people, genes and properties, goods and attachments, in brief *cosmograms*."[3] In this sense, politics is a process of building relations between humans but also between humans and nonhumans. Embracing such an expansive notion of politics to focus on the interactions between human and nonhuman actors allows us to interpret and act upon urban nature in new ways.

I begin the chapter by describing the dominant form of environmental politics in Austin and Seattle, what I term "rational politics." Here, the municipal government is the main arbiter of human/nonhuman relations, an evolution of the Promethean Project and technomanagerial governance as described in chapter 1. Rational political actions are challenged by grass-roots activities undertaken by individuals and local organizations, what I term "populist politics." These activities expose the deficiencies in techno-managerial governance and challenge rational actors to reform their activities and forge new relations. And a third form of politics, "civic politics," cuts across this top-down/bottom-up dichotomy of rational and populist politics by forwarding local, deliberative, and action-oriented programs for reworking urban nature. Civic politics has the potential to enact the relational perspective of urban nature as described in chapter 2 while reinventing the roles of urban residents, governments, and technical experts through radically different forms of political life.

"Leave It to the Experts": Rational Politics

Like all cities in developed countries, the most recognized and legitimate form of environmental management in Austin and Seattle is dominated by rational political actors. The emphasis of rational politics is on the development and upkeep of the nexus of technical experts, the state, legislation, and infrastructure networks. Rational politics in the United States emerged in the late nineteenth century with the Sanitary Movement and gradually transformed into technomanagerial governance. During this same period, there was a definitive split over the human relationship with nature. John Muir argued for a preservationist approach to environmental management to protect pristine natural areas from human destruction; Gifford Pinchot developed the conservation ideology of Wise Use to professionally manage natural resources effectively and efficiently. Despite their markedly different perspectives, they shared a faith in top-down, technomanagerial governance to enact the dominant conception of nature/culture relations.[4] Today, rational politics is understood as the most effective approach to mediate human/nature relations through state-sponsored practices of regulation, enforcement, and oversight.

In Austin and Seattle, rational politics of urban runoff is practiced by the municipalities who are charged with building, maintaining, and upgrading the stormwater networks. The regulatory approach of the City of Austin's Watershed Protection and Development Review Department and Seattle Public Utilities enrolls scientists, engineers, contractors, code writers, program managers, and inspectors to assess and monitor environmental conditions, create and enact municipal legislation, upgrade and extend the existing drainage networks, and enforce pollution regulations. Rational politics is typically described as a "command-and-control" form of environmental governance with technical experts as central actors.[5] This approach was particularly evident beginning in the late 1960s, as federal legislation such as the National Environmental Policy Act, the Clean Water Act, and the Clean Air Act solidified the federal government's role as the ultimate manager of environmental quality. The administrative state extended its purview beyond social regulation to include nonhumans, with the ultimate goal of establishing and enforcing socially acceptable levels of environmental pollution.

The criteria for rational governance are determined by engineers, scientists, and lawyers—a group of professionals that Shutkin refers to as the "holy trinity" of environmental expertise.[6] These experts are assumed to stand above the political process while the public benefits from the application of their specialized knowledge, carrying over the nineteenth century concept of apolitical expertise.[7] In the eyes of technomanagerial actors, politics is a problem rather than a solution to the nature/culture dilemma; thus, environmental pollution is most effectively addressed by individuals who can implement solutions free of unscientific bias.[8] This emphasis on knowledge production from above marginalizes alternative understandings of humans and nonhumans by forwarding a rigorous separation between facts and values.[9] Political scientists Karin Bäckstrand and Eva Lövbrand summarize this position when they state, "Through a detached and powerful view from above—a 'global gaze'—nature is approached as a terrestrial infrastructure subject to state protection, management and domination."[10]

Rational politics is most often dominated by government actors but also includes civil society groups. This was particularly evident in the 1980s as the so-called Big Ten environmental groups adopted legal and scientific approaches to combat an increasingly antienvironmental federal administration.[11] The rise of counterexperts generated alternative scientific and technical opinions on acceptable levels of environmental pollution and opened up environmental expertise to critique and debate. Reflecting on the emergence of counterexperts, political scientists Marc Landy, Marc Roberts, and Stephen Thomas write, "Supposedly neutral expertise becomes the weaponry of partisan conflict. The result is at best confusion about the appropriate role (and limits) of technical knowledge, and, at worst, the widespread belief that experts are mere hired guns who have nothing to contribute to the policy debate. When that happens, both government and ordinary citizens find it difficult to learn about a problem since they do not know who, if anyone, can be trusted."[12] Environmental groups in Austin and Seattle have engaged in practices of counter-expertise to redirect the technomanagerial approach to urban runoff. The work of the SOS Alliance in Austin and its success in passing the SOS Ordinance in 1992 is perhaps the most vibrant example of counterexperts challenging rational politics. SOS used existing avenues of governance—namely, regulation and litigation—to shift practices of urban development in a new direction.

The government/civil society dichotomy of rational politics that emerged over the last three decades mirrors the preservation/conservation dichotomy of the late nineteenth and early twentieth centuries in its embrace of technomanagerial governance.[13] There is an implicit understanding that this is the "proper" mode of environmental management, which means that rational political activities are limited to strengthening existing regulations or creating new regulations as well as updating bureaucratic procedures. There is no question that the state is the central arbiter of human/nature relations.[14]

One of the most notable deficiencies of rational environmental governance is its dispersed and fragmentary treatment of pollution, an approach that fails to acknowledge the complex character of environmental problems.[15] Dryzek argues that "the very idea of holistic pollution control flies in the face of Weberian administrative logic."[16] The bureaucratization of environmental management focuses on legislation to target single pollutants, discrete sources, and single media because they are the most accessible and soluble to government regulation.[17] In other words, rational governance of environmental pollution loses the forest for the trees; there is no systems view of environmental pollution but rather an emphasis on isolated parcels.[18] Sustainable development theorist Mark Roseland summarizes this critique: "Conventional wisdom considers the environment as an administrative problem, to be solved by better management—understood as cutting the environment into bite-sized pieces. The approach seems increasingly unable to deal effectively, sensitively, and comprehensively with environmental complexities."[19]

Using terminology from chapter 2, rational governance can be understood as forwarding a topographic view of the landscape, focusing on discrete pollutants and land use activities rather than recognizing the interconnectedness and complexity of human/nature relations. Furthermore, this topographic understanding of environmental pollution tends to emphasize pollution control over prevention.[20] The underlying assumption is that environmental degradation is an inevitable product of contemporary society and that the role of environmental experts is to establish and maintain socially acceptable levels of contamination. When pollution measures are put in place, these measures frequently involve shifting pollutants from one medium to another (e.g., water to air) or from one geographic locale to another, with low-income and minority communities often shouldering the majority of environmental risks.[21]

This is not to argue that rational politics is completely ineffective or that it should be abandoned outright. To be sure, regulation and the control of pollution has resulted in substantial gains in environmental quality and human health since the nineteenth century, particularly with respect to point sources of pollution and the so-called low-hanging fruit that can be addressed with relative ease via state-led bureaucratic management. However, the approach has been less effective with more complex environmental problems such as nonpoint source pollution and climate change in which dispersed sources confound compartmentalized bureaucratic regulations.

In cases in which problems cannot be simplified to technomanagerial governance, regulatory authorities resort to public education to persuade the public to act as more diligent environmental stewards.[22] With urban runoff, structural controls are those strategies that the municipal government can control, and nonstructural controls are those that are outside of technomanagerial governance, creating a division of labor between government and residents as well as a cognitive split between technical and social approaches to environmental management.[23] This two-pronged approach of rational politics, composed of prescriptive government activities and voluntary public actions, is the dominant form of environmental management practiced today and creates a particular arrangement of human/ nonhuman relations. Urban residents are understood as receivers of municipal environmental services rather than an integral part of the nature/ culture nexus and are called upon only when technomanagerial strategies cannot meet environmental quality goals.

Furthermore, community advocates argue that rational politics and technomanagerial governance strategies are insufficient to develop long-term, democratic forms of governance. Political scientist Frank Fischer makes an explicit connection between environmental flows and governance, arguing that "the environmental crisis is as much a crisis of the institutions that have to interpret and regulate risks as it is a physical phenomenon pertaining to natural processes."[24] This suggests that the solutions to the problem of environmental degradation in cities not only lies in the development and enforcement of more stringent environmental regulations but also involves a fundamental reassessment of the political institutions that conceptualize and manage urban nature.

"Power to the People": Populist Politics

The flip side of the top-down, broadly applied regulatory approach of rational politics is populist politics. This model, as its name implies, is bottom-up and is practiced by nongovernmental organizations and individuals who focus on discrete local issues. The environmental justice activities in East Austin and the Northgate Mall controversy in Seattle are both quintessential forms of populist politics directed at issues of urban runoff. In both cases, individuals and local organizations directly challenged the economic development strategies of the municipal government and powerful consortiums of property owners and developers due to the social and environmental impacts that would result. The populist actors highlighted the negative implications of applying citywide or regional development strategies to their local communities, widening the gulf between government and the people.

Populist politics is frequently contrasted with rational politics, but this is not to argue that it is an "irrational" form of politics. Rather, it presents an alternative to the expert-dominated, bureaucratic character of rational politics. Shutkin identifies Alice Hamilton, a pioneering urban environmentalist who studied industrial disease and occupational hazards, and Jane Addams, founder of Chicago's Hull House Settlement in 1888, as early practitioners of populist politics in the United States.[25] Hamilton and Addams championed grassroots activities to promote sanitary and public health reform on behalf of urban immigrant and low-income populations, but they tend to be marginalized in popular accounts of U.S. environmental history because their work falls outside of the dominant preservationist/conservationist debate of environmental management.[26] In short, their expansive notion of environment that included humans was in conflict with the scientific management approach of conservation and preservation advocates.

Today, the environmental justice movement serves as the most visible form of populist politics in the United States. Political scientist DeWitt John characterizes this approach as the conscience or emotive heart of environmentalism and traces a political lineage from Andrew Jackson to civil rights activist Rosa Parks and toxic chemicals activist Lois Gibbs.[27] Environmental justice proponents and other populist political actors critique rational politics by highlighting the suppressed voices of nonexperts, namely community residents, who are adversely affected by the policies produced through rational political activity.[28] Rather than developing counterexpertise, they call for more democratic forms of policy making to shift knowledge generation from the realms of science, engineering, and economics to specific temporal and material contexts. As such, the universal, general, and timeless approach of rational politics is replaced with a political program that is particular, local, and timely.[29] Political scientist Margaret Canovan writes, "Populists see themselves as true democrats, voicing popular grievances and opinions systematically ignored by governments, mainstream parties and the media. Many of them favour 'direct democracy'—political decision making by referendum and popular initiative. Their professed aim is to cash in on democracy's promise of power to the people."[30] Opening up environmental politics to more voices is accomplished by acknowledging the importance of local experts, those residents with intimate knowledge of local problems that are frequently overlooked in expert debates about environmental legislation.[31]

Ironically, while populist politics are frequently anti-governmental, they often stimulate the creation of new rational legislation aimed at righting the social wrongs created by earlier legislation.[32] A feedback loop between rational and populist environmental governance emerges to correct for intentional as well as unintended effects of top-down environmental regulation. This feedback loop of rational/populist politics is evident in both the Austin and Seattle case studies. Northgate community activists translated their populist political energies into rational environmental legislation through the Yes for Seattle campaign, an approach that ultimately failed due to legal and legislative barriers. And the protest activities of East Austin community members succeeded in shifting municipal policy to address the uneven environmental and social conditions in their community.

Similar to rational politics, populist politics is an important approach to reworking urban nature. Populist political activities challenge techno-managerial governance by forwarding alternative understandings of nature/culture conflicts that are not considered by rational political actors. The inclusion of alternative perspectives is ultimately a call for more democratic forms of politics that reject the apolitical stance of technomanagerial governance. Unlike rational politics, the approach understands human populations as embedded in their material surroundings and recognizes that human/nature relations need to be a part of environmental decision-making processes.

However, populist politics is similar to rational politics because it embraces a topographic perspective of environmental problems, recognizing human/nature relations in particular locales, but fails to connect the local perspective with other parts of the city. Whereas rational politics reinforces the dichotomy of humans and nature, populist politics sacrifices the broader purview of technomanagerial governance for a decidedly local perspective. East Austin activists addressed the symptoms of uneven urban development in their neighborhoods but did not tackle the larger problems of urban growth, economic inequity, and other similar issues that created these symptoms in the city as a whole. Likewise, the Northgate controversy resulted in a design solution to ease the tensions between local stakeholders but did not connect this work with either upstream or downstream portions of Thornton Creek or other Seattle creeks. The outcomes of these projects will arguably benefit the specific locales in question, but they do little to address similar issues in other neighborhoods.

In addition to its topographic interpretation of environmental management, a significant drawback of populist politics is that it often fails to produce long-lasting changes to environmental governance. Environmental justice scholar Julian Agyeman writes, "Grassroots environmental justice groups are often lacking in their ability to frame the issue, seize on political opportunities, and mobilize the political and financial resources needed to be more proactive, that is, heading off problems before they arise."[33] In other words, populist politics tends to be reactionary, leaving the larger structure of technomanagerial governance in place.[34] Like rational politics, there is no questioning of the state as the ultimate manager of human/ nature relations.

So is there a way out of this feedback loop of rational and populist politics? Is this tension between government and the people the "natural" and unavoidable reality of contemporary politics? Or is there the potential for what Gandy refers to as "a new kind of environmental politics that can respond to the co-evolutionary dynamics of social and bio-physical systems"?[35] Can a relational perspective of urban nature be transformed into constructive action?

Building Relations, Finding Society: Civic Politics as Topological Practice

The U.S. environmental legislation of the 1970s is often heralded as a triumph of the environmental movement to institutionalize environmental protection and has undoubtedly led to significant improvement in environmental conditions, particularly for easy-to-control pollution problems from industrial and municipal point sources. However, the 1980s represented a new era of federal environmental politics, as the conservative administration of Ronald Reagan attempted to weaken and overturn 1970s environmental regulations and the U.S. Congress was mired in deadlock over the promulgation of new environmental regulations. The U.S. Environmental Protection Agency was established as a separate governmental unit from the natural resources units of the federal government such as the U.S. Army Corps of Engineers, the Bureau of Land Management, and the Bureau of Reclamation; environmental protection was thus not integral to all government agencies but was instead yet another layer of federal technomanagerial regulation.[36] During this era, it became commonplace to pit polluters against protectors of the environment and to frame environmental protection as a zero-sum game of jobs versus environmental protection.[37]

Amid this dichotomous political atmosphere of the 1980s, a new form of environmental politics began to take shape. The term "civic environmentalism" was coined by political scientist DeWitt John to refer to innovative approaches to environmental politics and governance characterized by more flexible, collaborative solutions to complement, rather than re-place, the top-down, rational approach to environmental problem solving that has dominated U.S. politics since the late nineteenth century.[38] John writes:

Civic environmentalism is the process of custom designing answers to local en-vironmental problems. It takes place when a critical mass of community leaders, local activists, and businesspersons work with frontline staff of federal and state agencies and perhaps with others to address local issues that they care about deeply. Civic environmentalism cannot succeed without some participation and support by government agencies, but it is essentially a bottom-up process that epitomizes reformers' aims to build a results-based sense of common purpose in environmental governance.[39]

John's interpretation of civic environmentalism tends to emphasize the centrality of state governments in environmental policy making rather than the federal government and is exemplified by projects such as the Chesa-peake Bay Program, the Florida Everglades restoration, and the reduction of agricultural chemical use in Iowa.[40] These are noteworthy projects that include "economic incentives, technical assistance, public education, and voluntary government programs, all tailored to local conditions" as well as collaboration between experts in business, government, higher educa-tion, and nonprofit organizations.[41] John characterizes these approaches as customized, local approaches to environmental governance that are akin to the "third way" or "radical center" political program promoted in the 1990s by centrist political leaders such as President Bill Clinton and UK Prime Minister Tony Blair.[42] Using a carrot-and-stick approach that mixes market and interventionist governance, proponents of Third Way politics call for the development of savvy strategies to overcome the liberal/conservative dichotomy of contemporary politics in North America and Northern Europe.[43]

John's civic environmentalism calls for the continued emphasis on envi-ronmental governance informed by ecological science but at a smaller scale in which more nuanced regulatory and nonregulatory strategies can be enacted, particularly through interstate public/private partnerships. Such an approach opens up environmental policy making to more stakeholders but continues to focus on technomanagerial governance. Indeed, John is at pains to define civic environmentalism as a "complement" to rather than a "replacement" of conventional environmental management.[44]

However, others have interpreted civic environmentalism as a potential way to transform populist political activity into constructive local action. Scholars and activists of sustainable community development recognize civic environmentalism as a way to channel the energies of environmental justice and related activities into new forms of political organization and effectively go beyond the governmental reform agenda of John's civic environmentalism.[45] From their perspective, civic environmentalism can do more than reform existing modes of government and can instead create entirely new modes of collaborative problem solving at the community level, the most basic level of political organization.[46] The term "civic environmentalism" is thus a misnomer because it is not an anthropocentric form of environmentalism but rather a *politics of urban relations* that recognizes the interdependence of social, environmental, and economic problems. These practices of relation building, what I call "civic politics," engage urban residents at the local level and replace the protest activities of populist politics with deliberation and action aimed at creating and maintaining more desirable conditions.

Civic politics not only provides an alternative to rational and populist politics, but also serves as a way to *enact* the relational perspective as described in chapter 2. In his 2004 book *Politics of Nature*, Latour argues that the process of assembling networks is what should be rightly called politics. The common work of politics and the sciences is to stir "the entities of the collective together in order to make them articulable and *to make them speak*."[47] Latour argues that a relational approach suggests a very different form of public organization, one that considers the material and nonmaterial simultaneously. Geographer Noel Castree critiques Latour's notion of politics, arguing that Latour "has yet to flesh out precisely what 'politics' might mean in a world of [relations]," adding, "It is one thing to have a new political vocabulary, but quite another to have substantive political concepts that ground new forms of practice."[48] Geographer Jonathan Murdoch makes a similar critique, arguing that "Latour's work on political ecology is strong on *re-conceptualizations* of science and politics in the wake of political ecology but is weak on the specific steps that might be taken to shift scientific and political *practices* in the desired direction."[49] There is a significant gap between the relational theory of Latour (and likeminded human and cultural geographers) and any form of political structure for collective action. Likewise, landscape architects, ecological planners, and other designers who practice relation building frequently do so on a site-by-site basis without scaling up their approaches to a new political program of urban nature. This is why Gandy calls for a new kind of environmental politics—a program that can transform relational thinking and practice into a transformative political agenda.

Civic politics provides relational theorists and practitioners with a political program to *enact* a relational perspective by encouraging different *thinking* about human/nonhuman relations and, more important, different *doing*. It is here where relational perspectives on urban nature take on their muscle and worth to society rather than merely serving as a theoretical construct or a one-off intervention. It is through the application of this knowledge in the reinvention of political life that the relational perspective of urban nature becomes useful.

Like populist politics, civic politics recognizes that humans are embedded in their material surroundings. The physical world is not simply a platform for practicing environmental politics; it is co-constitutive in the practice of understanding and reorienting ecological flows.[50] The particulars of place are championed over universal interpretations of an issue because it is at the local scale, in which those who are most affected can identify the most important problems. Rubin argues that "people care passionately about what is close to them,"[51] and John adds that civic politics involves "a deep, shared commitment to a physical place and to the community of people who live there."[52] In short, civic politics takes to heart the 1970s environmental mantra of "think globally, act locally."[53]

This emphasis on the local is divisive because technomanagerial governance frequently champions the importance of the global scale while situating the local at the bottom of the bureaucratic hierarchy. Political scientist Timothy Luke writes, "Local knowledges, vernacular technics, and civic sciences as environmental mediations . . . are dismissed before the privileging of international knowledge formations, transnational technical networks, and national scientific societies that embrace globality."[54] Civic politics challenges the primacy of the global as the fundamental scale of environmental politics and management; the local is not the *only* scale of environmental politics but it is recognized as the most significant. This emphasis on the local is an attempt to rescue the maligned term "community" from nostalgic, conservative interpretations; here, community is understood as the core of democratic politics.[55] Shutkin summarizes this emphasis on a local politics of place: "In a civic democracy, place and community are mutually constitutive and reinforcing."[56] The emphasis on place and community reverberates with philosopher John Dewey's attempts to define the public through local democratic action. He sums up this position, stating that "democracy must begin at home, and its home is the neighborly community."[57]

This grounding in place involves multifaceted relations and requires a pluralist interpretation of urban nature that recognizes the validity of multiple perspectives and experiences. Civic politics recognizes that "there are many different kinds of environmental problems and different points of view."[58] Political scientist Lamont Hempel notes:

Partly in response to dissatisfaction with the sustainable development concept and partly in response to growing concerns about urban quality of life, a splinter movement of sorts has arisen in an effort to focus sustainability strategies on the social, economic, and ecological well-being of communities. Participants in this movement define community sustainability in ways that highlight the relationships between local quality of life and local or regional levels of population, consumption, political participation, and commitment to intertemporal equity.[59]

Civic political actors operate in neighborhoods to organize diverse coalitions of organizations and activists who address the overlapping issues of public health, growth management, urban planning and development, environmental justice, urban environmental quality, and local government control. The local scale is not just a material place where relations between humans and nonhumans are readily apparent but also a place where these relations can be deliberated and acted upon.[60] It is at the local scale where urban residents recognize that "environmental quality and social health are inextricably linked."[61]

Civic politics is not aimed at decentralizing environmental policy making but rather at creating a transformative mode of local politics steeped in deliberative democracy and community activity. Such an interpretation of civic politics is founded on the belief that sustainable urban development is a deeply political project and "sustainability will be achieved, if [at] all, not by engineers, agronomists, economists and biotechnicians but by citizens."[62] This belief resonates with the various calls by political science scholars for strong democracy, deliberative democracy, and ecological democracy that emerged in the mid-1980s.[63] Building on historical figures such as Thomas Jefferson and Alexis de Tocqueville, as well as more contemporary philosophers such as John Dewey, John Rawls, and Jürgen Habermas, the emphasis on deliberation in democratic process rejects aggregative or representative models of democracy and instead, recognizes the link between collaborative problem solving and community building.[64] The importance of democracy is not in its prescribed models but in the processes of deliberation, reasoned argument, and public reflection.[65] Dryzek argues that "the main reason for the democratization of environmental administration has been a felt need to secure legitimacy for decisions by involving a broader public."[66]

There is no guarantee that deliberative democratic procedures will result in more ecologically oriented communities, but it has great potential to resolve environmental problems by providing a process for integrating a variety of perspectives on complex human/nature relations. Dryzek writes, "Discursive democracy is better-placed than any alternative political model to enter into fruitful engagement with natural systems and so able to cope more effectively with the challenge presented by ecological crisis."[67] In other words, deliberative forms of politics at the local level have the most promise for recognizing and reworking complex human/nonhuman relations and for securing legitimacy of decisions by involving the broader public. The governance of human/nature relations is not dominated by a unitary state administration composed of experts but rather a multiplicity of active citizens who are deeply engaged in decision-making processes. With civic politics, "*Homo civicus* figures large, *homo bureaucratis* hardly at all."[68] There is an insistence on urban residents to treat citizenship as a serious vocation and to engage in political debate and decision making based not on a founding environmental ethic or a commitment to the state but rather a responsibility stemming from their embeddedness in place.[69]

However, civic politics diverges from deliberative and discursive democracy because of its emphasis on materiality. Deliberation is not only comprised of social interaction but must account for and engage with the physicality of place. Further, democratic deliberation is not only a means of fleshing out the commonalities and differences between urban residents but is also intended to catalyze constructive action. The activity of democratic deliberation is not intended as idle talk or just another planning exercise, nor is it intended to reform rational political activity. Rather, it is a means to develop and fundamentally change existing forms of local politics as well as urban governance. Civic politics is aimed at "tapping into the public's pent-up demand for effective, hands-on community-building strategies" and "intentionally incorporates the notion of praxis: the conjoining of social and political ideas with new social practices and technologies."[70]

Civic politics starts with a local or regional problem, similar to populist politics, and then develops a comprehensive, far-reaching solution, similar to rational politics. However, the emphasis is not on the problem definition or on the ultimate solution but rather on the process of *developing* the problem definition and solution. As such, it involves a pragmatist emphasis on problem solving rather than developing ultimate regulations or an idealized community as an end product.[71] These activities foster what philosopher Andrew Light calls "urban ecological citizenship," the engagement of the individual in the maintenance of natural processes in cities.[72] It is through activity with our surroundings, both human and nonhuman, that we become politically engaged actors, or citizens, who enact human/nonhuman relations.

Civic Politics and Urban Runoff

With respect to urban runoff, rational politics has dominated over the last century because the governance of urban water has been the responsibility of the state. Populist political activities arise less frequently and tend to center on human health issues related to rational political activities such as flooding incidents that endanger human lives and property. And civic political activities almost never focus on urban runoff issues.[73] However, there is great potential for urban water activities to be a central element of civic politics, as illustrated by landscape architect Anne Spirn's work in the Mill Creek neighborhood of West Philadelphia.[74] Spirn first began working in this blighted, low-income, African American neighborhood in 1987 because she was interested in using vacant lots to simultaneously restore urban nature and improve local conditions for residents. Her investigation shifted to urban runoff issues when she realized that the empty lots correlated with a historic waterway, Mill Creek, that had been filled in and paved over through urban development processes. It gradually dawned on Spirn that these residents lived at the bottom of a watershed as well as the bottom of the social economy, a common characteristic in Austin and Seattle as well as many other communities.

Rather than proposing an expert-dominated, environmental restoration project to resolve the conflicts between social and environmental conditions, Spirn broadened the project to include community members and her students at the University of Pennsylvania in hands-on activities to transform some of the low-lying lots into outdoor classrooms, demonstration projects, detention facilities, and community gardens.[75] Spirn writes, "To read landscape is also to anticipate the possible, to envision, choose and shape the future: to see, for example, the connections between buried, sewered stream, vacant land and polluted river, and to imagine rebuilding a community while purifying its water."[76] Spirn's work on Mill Creek is a combination of historical research, community engagement, collaborative design, and action-oriented pedagogy. It is also an example of the transformation of Spirn from a landscape architect who focused on form and a topographic understanding of landscape to a politically engaged actor with a topologic perspective emphasizing contextual processes that enact the hybrid urban landscape. Her work is simultaneously cultural, natural, and technical, with community action resulting in a new form of relational politics that is distinct from conventional landscape architecture practice.

In Austin, civic politics related to urban runoff are nascent at best. One of the most fully developed examples is the Country Club Creek Trail, a project birthed from a municipality-led neighborhood planning project in the early 2000s. Building on the momentum of the rational planning activities, a small group of residents in the East Riverside and Oltorf neighborhoods of southeast Austin began meeting in 2004 to plan and construct a hike-and-bike trail that would follow Country Club Creek and connect the green spaces in their neighborhood. The group received donations of materials and money from area businesses, the Austin Parks Foundation, and Keep Austin Beautiful, and as of January 2011, about one mile of trail had been built by volunteer labor. The municipal government has maintained a limited role in the trail project, helping to acquire a recreational easement for the project and to remove trash collected by volunteers.[77] At this point, the project is simply a trail, but it has the potential to open this neglected urban space to alternative notions of human/nature relations while engaging residents in political activities that are deliberative and constructive.

In contrast to Austin, Seattle offers several examples of community-based projects that forward a civic politics of urban runoff. Two of the most notable of these projects are the Longfellow Creek Legacy Trail and Growing Vine Street, both of which are described in detail in the following paragraphs. These projects demonstrate the potential for urban water to catalyze new forms of political organization that embrace a relational perspective of urban nature while involving very different forms of citizenship, governance, and expertise.

Since the early twentieth century, Longfellow Creek has served as the primary spine for urban drainage in the Delridge neighborhood of West Seattle. After World War II, the stream channel was straightened and armored to make way for single-family residential development; today, the overall impervious surface coverage of the watershed is 52 percent.[78] About half of the watershed is served by a combined sewer network and the other half consists of separated or partially separated sewer networks and informal ditch-and-culvert networks (the latter two networks discharge directly into the creek). There is little high-quality instream habitat in Longfellow Creek due to encroachment of development on the riparian zones, and significant portions of the creek have been routed through underground pipes, including the headwaters (to make way for a shopping center in the 1950s) and the mouth of the creek (to accommodate industrial and transportation development in the 1960s and 1970s). Today, the creek is a tortuous route for migrating salmon, with seven complete fish barriers and three partial fish barriers.[79]

Flooding incidents are common in the low-lying areas of the watershed during large storm events due to impervious surface coverage and the close proximity of residential and commercial properties to the creek channel. In the late 1980s, the Delridge Community Association began a formal process with the City of Seattle to prepare an action plan to identify and prioritize development issues in the community, and drainage issues emerged as the highest priority. The City of Seattle partnered with King County and the State of Washington to purchase and preserve thirty acres of property adjacent to the creek to reduce flooding problems.[80] In many cases, this process involved buying houses and demolishing them to make room for the creek—an expensive and intrusive redevelopment practice.

In 2000, Delridge neighborhood activists devised a plan to create a trail along the creek that would simultaneously restore habitat and encourage resident interaction with the waterway. The trail was intended to fulfill the first strategy of the Delridge Neighborhood Plan to "integrate the community with nature."[81] The residents received grant funding from the City of Seattle to build the Longfellow Creek Legacy Trail, a 4.2-mile trail along the creek that would serve as a "ribbon of connection" for the community.[82] The trail was completed in 2005, with related volunteer and municipal government activities resulting in the restoration of nearly 40 percent of the open creek channel.[83] The interrelated goals of these activities have been to improve water quality and drainage conditions, prevent erosion and flooding, restore habitat, and expand community open space and trails.

The Longfellow Creek Legacy Trail begins at a ten-thousand-year-old restored peat bog and follows the creek until it goes into an underground pipe to traverse the aforementioned shopping center. The trail continues on the surface following a virtual creek marked by signage and a water feature in the shopping center that reminds shoppers of the waterway flowing some twenty feet below the surface. Downstream from the shopping center, the creek surfaces again and the trail follows along, sometimes adjacent to the waterway and other times diverted along residential streets where access to the creek is impossible. In extreme cases, the trail jogs two blocks from the open channel and runs parallel down city sidewalks before meeting up with the creek downstream. Local writer Kathryn True writes, "The Legacy Trail is an exercise in contrasts. From lush, shady areas that quietly conjure a mountain stream, to sections that border convenience stores and the rush of traffic."[84] The trail ends unceremoniously about two-thirds of a mile before reaching the Duwamish River, when the creek again goes into a pipe to traverse a steel plant and a knotty mess of transportation infrastructure.

The trail is not intended as a traditional nature walk or a recreation venue for local residents but as a way to experience the hybridity of the city by following the path of water from its headwaters to the Duwamish River. It is the material realization of Janice Krenmayr's experiential urban hikes of the 1970s as described in chapter 6 and forwards human experience as central to the reworking of the urban landscape. In addition to wayfinding signs and gateways, the trail is peppered with creative art pieces, such as the Salmon Bone Bridge and the Dragonfly Pavilion, that serve as educational tools and visual metaphors for the creek (see figure 7.1). Reflecting on environmental art as it relates to place, writer Lucy Lippard notes, "Artists can be very good at exposing the layers of emotional and aesthetic resonance in our relationships to place. . . . A place-specific art offers tantalizing glimpses of new ways to enter everyday life."[85] This

Figure 7.1
Elements of the Longfellow Creek Legacy Trail, including the water feature in the Westwood Shopping Center (left), a wayfinding sign (top right), and the Salmon Bone Bridge (bottom right).

process of exposure opens the imagination of residents to the hybrid relations that constitute the city. However, it comes at a cost; while exposing the messy relations between human and nonhumans along the creek, the Longfellow Creek Legacy Trail has had unintended impacts on the biological and physical health of the creek. An SPU staff member notes, "In some places, the trail is right beside the creek and the stormwater runoff from the trail goes right into the water. In other places, riparian zones that are crucial for salmon habitat have been removed to make way for the trail. And there are great opportunities to daylight the creek and link it to the existing floodplain but the trail goes right between them."[86]

There is a shared understanding by the neighborhood residents that they are not trying to return the creek to predevelopment conditions but rather to use the creek as a cognitive and physical connection between residents and their surroundings. A Seattle Parks and Recreation Department staff member sums up this position, stating that "common sense guides restoration [projects]. In an urban area, restoration is going to be a balance between the built environment and what you can achieve to create connection and bring back the [preexisting] processes."[87] There is an emphasis on process rather than product, and activities of negotiation, mutual learning, consultation, and hands-on work by community members. A community member active in the trail planning deliberations states, "I was impressed when we held . . . community meetings that there were at least 50 people at each meeting. That was a good indication [that the community was interested in the project]. Two or three people at each meeting would be really upset, but 99 percent of the people there were really excited and wanted to see it happen."[88] Overall, this approach leads to the realization that the work on Longfellow Creek is as much social as it is biological.[89] There is clearly an interest in restoring biological habitat, but this interest is balanced with the experiential aspects of the trail. Urban water is interpreted here as an opportunity to reveal and experience nature while building community.

Many urban drainage activities in Seattle focus on neighborhood creek projects in residential areas because of the visual and physical proximity to urban inhabitants. However, urban water activities also occur in highly urbanized areas where the original waterways are no longer present. These projects often piggyback on another form of urban nature: community gardens. Seattle's P-Patch or community gardening program started in the early 1970s as part of the back-to-the-land movement.[90] In the intervening decades, the program has grown to include over fifty gardens serving seventy neighborhoods and six thousand urban gardeners on twenty-three acres of land; today, it is the largest municipally run community gardening program in the United States.[91]

An exemplary urban runoff project attached to a community garden is Growing Vine Street, just north of the central business district and next to the Belltown P-Patch. Since the mid-1970s, environmental artist Buster Simpson has used the Belltown neighborhood as his canvas, creating projects to inspire urban residents to take responsibility for their surroundings. He has been described as an "agent provocateur" or "trickster" because he opens up the city to environmental flows that urban residents often fail to acknowledge.[92] Simpson's political outlook is heavily influenced by Seattle's participatory turn in local governance during the 1970s and this outlook is embodied in his art projects. Reflecting on the underlying motivation for his work, Simpson writes, "Interventions and temporary prototypes provide a visible and engaging presence for ideas. This helps keep the community engaged for a time when collective consensus is needed to support more ambitious projects."[93]

In the 1980s, Simpson began to collaborate with Belltown community activists to preserve their community garden space and a series of historic cottages on an adjacent property. Part of this work involved extending the community garden out onto Vine Street, a sloping downtown street that looks out on Puget Sound. Simpson found an ally in Carolyn Geise, a local architect who owns a historic building on the street and together, they championed the Growing Vine Street project in 1996. Simpson describes the approach to the project as: "defining the word 'green' in relation to environmental sustainability rather than to traditional landscaping. This part of the city was plumbed to dispose of rain from roofs and hard surfaces through an antiquated combined sanitary and storm system. We proposed to redirect this urban watershed and keep it at the surface as an asset rather than a liability to be flushed out of sight. We have put the city on notice that we see gray water and brown water as the next opportunity."[94]

Simpson and Geise framed Growing Vine Street as an urban laboratory for stormwater solutions in which greening of urban infrastructure could be accomplished in conjunction with fostering local community cohesion. Reflecting on their design philosophy, Geise writes, "Our motto was, store the water, enjoy and play with the water, irrigate with the water; do not just send it down a black hole to get rid of it."[95] Likewise, Simpson writes of creating a "crack" or "fissure" in the impervious Belltown landscape to bring stormwater flows back to the surface.[96] He writes, "Rather than fighting the infrastructure, we let it reveal itself by working with the existing conditions and taking the path of least resistance."[97]

The most visible element of Growing Vine Street is the Beckoning Cistern, a ten-foot-tall, six-foot-diameter blue steel tank that tilts toward Geise's historic building. The cistern is a hybrid of sculpture and infrastructure, with "fingers" that reach out from the top to a downspout on the building, mimicking the hand of Adam reaching out to God on the ceiling of the Sistine Chapel.[98] The building's occupants use the rainwater collected in the cistern for their rooftop gardens. An earlier project on the same historic building is the Downspout-Plant Life Monitoring System constructed by Simpson in 1978. The existing downspout from the building was inverted to create planters for native ferns, resulting in a vertical landscape where none existed. One block closer to Puget Sound, a more traditional stormwater project called Cistern Steps was constructed in 2003, extending the P-Patch into the street with a set of marshy steps (figure 7.2). Not surprisingly, there were difficulties in navigating the complex layers of municipal codes and regulations, but the team had an ally in Mayor Schell, who felt that the neighborhood should receive amenities because it was accepting increased urban density to comply with the state's Growth Management Act requirements. The municipal government contributed $800,000 to the Growing Vine Street project and was assisted by a volunteer team composed of artists, design professionals, developers, and community businesses.[99]

Like the Longfellow Creek Legacy Trail, the Growing Vine Street project follows an ecorevelatory design strategy, using hydrologic flows to bring community members together and to reveal the presence of nature in the city. However, the highly urban context of Growing Vine Street was devoid of predevelopment drainage patterns; thus, the design team was given a clean slate on which to design new, albeit artificial, drainage patterns. They understood that there was no possibility of salmon populations ever repopulating the Belltown landscape and their project would have only minor impacts on the drainage patterns in the city. The logic was therefore aimed at social learning and heightened awareness of nonhuman flows in a part of the city that is devoid of unbuilt conditions, following on the goals of the Urban Creeks Legacy Program developed by Mayor Schell and the municipal government. The Beckoning Cistern and Cistern Steps also reinterpreted the public right-of-way to be more than driving lanes, parking spaces, and sidewalks; it was a place for public interaction and exploration rather than automobiles.

The aforementioned neighborhood-based drainage activities can be understood as different forms of civic politics, with local residents using environmental restoration as a means to emphasize different relational

Figure 7.2
The Growing Vine Street project includes the Beckoning Cistern (top left), the Downspout-Plant Life Monitoring System (top right), and the Cistern Steps (bottom).

aspects of urban water. These activities demonstrate a distinct break from the conveyance logic of conventional stormwater management, the reform efforts of Low Impact Development and source control advocates, and the populist protest of neighborhood residents. There is an emphasis on the experiential, biological, playful, artistic, and community-building opportunities offered by local water flows. A distinct feature of these projects is that they were directed and implemented by local community and environmental organizations. Various government entities were involved but generally served as funding sources and a source of expertise on a limited scale.

The Challenges of Civic Politics

To be sure, civic politics presents a number of significant challenges. Civic politics does not assume that deliberation and practice are simple tasks or that they can be conducted without deep disagreement and potential failure. Indeed, the politics described here are often more difficult to navigate than those of rational and populist politics because the ground rules have been significantly altered. One of the most damning critique of civic politics is that existing political structures are simply too entrenched, making radical transformation impossible. Commenting on the challenges of deliberative democracy, political scientist James Meadowcroft writes, "The adversarial political culture, legalistic regulatory approach, litigious proclivity, and deep suspicion of government found in the United States may represent insuperable barriers to the growth of this mode of environmental governance."[100] Furthermore, conservative political actors might actually be attracted to civic politics because of its potential to dilute environmental regulation or dispense with it altogether.[101] Thus, there is a worry that the deliberative process can be used to steer a community away from prescribed goals by replacing unsatisfactory regulations with no regulation whatsoever. Indeed, democratic deliberation does not automatically change the ground rules for debate; existing power geometries and power plays can still be present.[102]

These critiques represent a fundamental challenge to developing civic politics as a viable environmental governance strategy. However, the work of sustainable community activists and civic environmentalists suggests that radical political reorientation is possible. They counter the claim that contemporary politics cannot be rescued from its emphasis on intractable differences and have developed political processes in which participants take one another's positions and claims seriously.[103] Here, deliberation is not a contest between interests but an inquiry into solving problems through education, debate, and participation.[104]

Of course, reorienting environmental governance toward deliberative models requires time and energy. The purported advantage of rational politics is its efficiency in problem solving; the counterargument of deliberative democracy advocates is that this efficiency comes at a cost. Rational politics may be effective at addressing point sources (e.g., an industrial smokestack), but it often fails to address more systemic problems of nonpoint source pollution such as urban runoff and climate change. With problems that pose imminent danger, it may be necessary to curtail deliberative activities if a viable strategy of action cannot be devised in a reasonable period of time. But for most environmental issues, civic political actors argue that it is *more* democracy rather than less that will ultimately lead to more effective solutions to human/nature conflicts.

Another significant challenge to civic politics is that it calls for new modes of expertise, presenting a direct challenge to the top-down, techno-managerial forms of environmental governance. Proponents of deliberative democracy and civic environmentalism advocate for drawing on the local knowledge of nonexperts to complement formal expertise in environmental problem-solving activities.[105] Formal expertise is not to be abandoned but rather integrated in new modes of deliberation in which expert knowledge *informs* rather than dominates environmental problem solving. Professional and technical experts such as landscape architects, ecologists, and urban planners are used as consulting resources for reworking human/nonhuman relations and can provide a palette of options and scenarios for how urban nature relations could potentially evolve in space and over time. Further, their expertise can be used to facilitate and mediate deliberations among stakeholders about different configurations of human/nonhuman relations, interpreting complex issues to inform rather than dominate these activities.[106]

The insight of formal experts is supplemented by the specialized or expert citizen who contributes to deliberative and problem-solving activities through localized understandings and insights into their specific social and material surroundings.[107] Deliberative democratic practices thus offer a way to bridge the gap between expert and citizen by engaging ordinary people and experts in the collective project of addressing environmental and social problems.[108] The inclusion of nonexperts follows on Dewey's famous axiom, "The man who wears the shoe knows best that it pinches and where it pinches, even if the expert shoemaker is the best judge of how the trouble is to be remedied."[109] At the same time, citizenship comes with responsibility. Local experts are obliged to look outside their deep-seated beliefs about nature, politics, and community to engage in reasoned debate with uncertain and sometimes undesirable outcomes. It calls for faith in democratic processes to resolve human/nonhuman and human/nonhuman conflicts amicably. This is a tall order, to be sure, and calls for a political program that involves collaboration, compromise, and a shared goal of problem solving. Clearly, there is an idealistic tendency in discussions of deliberative democracy to treat politics as merely a process of negotiation.[110]

A final critique of civic politics is that, like populist politics, it cannot address problems that transcend local jurisdictions. Dryzek writes, "The larger the scale at which an issue arises, the harder it is to introduce discursive designs to resolve the issue."[111] It is arguably at these larger regional, national, and international scales that representative forms of politics have typically had the most success. However, deliberative democracy does not require that every actor be involved in every deliberation.[112] Rather, there is the possibility for representation through public enquiries, citizen juries, consensus conferences, and other forms of deliberation that are hybrids of representative and deliberative political structures.[113] Political scientist Richard Sclove writes, "What matters democratically is not the number of people but their interrelations and activity."[114] In other words, deliberative democratic politics can be scaled up to address larger problems by instituting different modes of representation and participation. In Austin, the Envision Central Texas project is aimed at developing a comprehensive regional strategy to accommodate population growth while improving economic, environmental, and social conditions.[115] In Seattle, the Open Space Seattle 2100 project was initiated by faculty members in the University of Washington's Landscape Architecture program. The project used the centennial of the original park system as an opportunity to update the open space plan of Seattle for the next century.[116]

Despite these challenges, civic politics offers a promising agenda for enacting a relational perspective of urban nature by fundamentally changing the structure of urban political activity. Civic politics blurs processes of deliberation and action as well as the boundaries between expert and nonexpert, public and private, citizen and government. The examples of the Longfellow Creek Legacy Trail and the Growing Vine Street project offer two examples of how urban nature can be reworked locally, actively, and collaboratively. These approaches are not based on a dogmatic insistence on the proper roles and responsibilities of government or an adherence to a particular environmental ethic but to the recognition that urban residents need to take their embeddedness in their human and nonhuman surroundings seriously and engage in deliberation and constructive action to continually rework and maintain their lived conditions.

Conclusions

The most dominant form of environmental activity today, rational politics, has made great strides in addressing the environmental impacts of industrialized societies. Its most prominent victories have come through the control of point sources, but this approach has also resulted in the bureaucratization of environmental protection. Populist politics serves as a check on rational approaches and opens up the tensions of nature and humans to more actors as well as issues of equity, justice, and context-specific problems. Both approaches are necessary but insufficient to address the complex issues of integrating humans and nature in urban contexts because they both recognize the state as the ultimate arbiter of human/nature relations, creating a division between humans and nature as well as government and citizens.

Civic politics, with its embrace of the complexities and nuances of human/nonhuman relations as well as its emphasis on engaged citizenship as

Table 7.1
A comparison of rational, populist, and civic politics of urban nature

	Rational	Populist	Civic
Problem-solving approach	Technocratic governance	Grassroots organization	Deliberative practice
Geographic focus	Federal, state	Site, neighborhood	Site, neighborhood, city, region
Primary activity	Rule making and enforcement	Community awareness and protest	Democratic deliberation and action
Key actor	Technical expert	Local activist	Engaged citizen
Austin examples	Watershed ordinances, conservation land development, SOS Ordinance	East Austin environmental justice	Country Club Creek Trail, Envision Central Texas
Seattle examples	Natural Drainage Systems approach	Northgate Mall redevelopment controversy	Longfellow Creek Legacy Trail, Growing Vine Street, Open Space Seattle 2100

the basis for constructive relation-building activities, provides an alternative model for reorienting the links between urban residents and their human and nonhuman neighbors. It takes the notions of hybridity, situatedness, and topology and transforms them into a political program of democratic deliberation and action. Civic politics requires us to be dogmatic not in our convictions to the environment, to society, or to government, but to the responsibilities we have as beings within a larger relational web. Although it is fraught with significant challenges, it provides an opportunity to rework the structural underpinnings of political life by applying relational thinking to real-world problems (see table 7.1).

If we are indeed awaiting a new form of environmental politics, as Gandy suggests, civic politics provides a promising—if challenging—way to reconceptualize what it means to live in the world and to rework human/nonhuman relations into more desirable forms. Changing contemporary modes of citizenship is not an easy task; indeed, this is a revolutionary rather than evolutionary prescription to address the dilemma of urban nature. However, there are incremental moves we can make to realize civic politics as the dominant mode of reworking urban nature. In the next chapter, I conclude by suggesting three modest proposals to nurture and develop this new politics of nature.

8

Toward the Relational City: Imaginaries, Expertise, Experiments

Urbanizing nature, though generally portrayed as a technological-engineering problem is, in fact, as much part of the politics of life as any other social process. The recognition of this political meaning of nature is essential if sustainability is to be combined with a just and empowering urban development; an urban development that returns the city and the city's environment to its citizens.
—Erik Swyngedouw[1]

The pursuit of sustainable cities is often portrayed as a technomanagerial endeavor to upgrade existing infrastructures to be more energy- and water-efficient; to develop regulations to reduce, reuse, and recycle materials; to foster more local economic activity; and so on. When we go beyond conventional interpretations of sustainability as a form of ecological modernization, it is clear that the pursuit of more sustainable urban futures has deep political implications that involve new modes of governance, citizenship, and daily life. The hybridity of the contemporary city requires a politics that can address the myriad economic, ecological, technological, and cultural connections between humans and nature, rather than an emphasis on upgrading and refining existing forms of technomanagerial governance. A civic politics of urban nature provides one route for acknowledging and acting upon the connections between humans and their material surroundings, and in the process, developing radical political programs for sustainable urban development.

In this chapter, I conclude by suggesting ways to foster more relational forms of civic political practice. Rather than propose a framework of deliberative democratic practice or prescribe best practices for managing urban runoff, I focus on three attitudes that are integral to realizing the relational city. First, there is a need to develop civic imaginaries that embrace the indelible connections between humans and nonhumans. Second, there is a need to reorient conventional modes of expertise so that those with specialized knowledge are tasked with facilitating rather than dominating political debates thereby opening up processes of urban development to alternative knowledges from informal local experts. Third, there is a need to trial new interventions in human-nature relations through practices of civic experimentation. Imaginaries, expertise, and experiments are aimed at catalyzing civic politics through discussion, debate, agenda setting, and action to realize new configurations of humans, nature, and technology in the city.

A Call for Civic Imaginaries

We need different ideas because we need different relationships.
—Raymond Williams[2]

One of the most daunting tasks of enacting civic politics is to replace the modern dichotomies of urban/rural, natural/artificial, human/nonhuman, and fact/value with a perspective that emphasizes the partial, hybrid, and messy connections of the world. This perspective has been and continues to be the main aim of relational theorists as described in chapter 2. One approach to nurturing a relational perspective and envisioning different futures is to forward the notion of ecological imaginaries.[3] The term "imaginary" is similar to common words such as imagination, image, and imagery, but is specifically focused on connecting vision with perception and meaning making.[4] Thus, an ecological imaginary is an attempt to bring into being new partnerships between humans and nonhumans, with each serving to reinforce the integrity of the other.[5] Ecological imaginaries are essential for all urban residents, experts and nonexperts alike, to form civic identity through the recognition of human/nonhuman relations.

Environmental philosopher Richard Dagger notes the etymological link between city and citizen, and argues that a *civic* imaginary allows the citizen to be not just *in* a city but also an integral part *of* it.[6] The citizen is one who is not merely a consumer, a producer, and a private individual, but is also linked in multiple and varying ways to his or her human and nonhuman surroundings. As such, ecological and civic imaginaries are closely linked; in both cases, the emphasis is on recognizing and nurturing constructive relations. A prescient example of a civic/ecological imaginary is expressed by a Seattle City Hall insider, an individual who does not self-identify as an environmentalist:

I remember going down to the Ballard Locks and watching the salmon coming through the fish ladder. And I realized, I live in a watershed, I don't live in a city!

It was the first time that I really understood that the core of the area is really around the watershed. The infrastructure is the watershed, not the roads and stuff. And when I tweaked my mind and thought about Lake Washington in a different way and thought of all of the channels and lakes and creeks, all of the sudden, that whole thing really changed the way that I thought about how we relate to our environment. And I think that happened to a lot of people, not necessarily in that big idea way but starting to look at our urban creeks and realizing that they aren't just pretty to have in your backyard. Or some of them that we would like to daylight and extend, there are reasons to do that besides the fact that this is aesthetic. And I think it gave more credibility to these creek restoration activities as we started to look at endangered species and habitat.[7]

The call for ecological imaginaries is not aimed at developing an environmental ethic in the traditional sense; rather, it is an attempt to elucidate the inextricable connections between humans and their surroundings and to recognize that this connectivity means that our actions always have intended and unintended implications. We do not merely live *in* a city, we are *of* the city; we are not merely *social*, we are *associated*. This reflects on Dewey's focus on the relational aspects of citizenship. He writes: "The characteristic of the public as a state springs from the fact that all modes of associated behavior may have extensive and enduring consequences which involve others beyond those directly engaged in them. When these consequences are in turn realized in thought and sentiment, recognition of them reacts to remake the conditions out of which they arose. Consequences have to be taken care of, looked out for."[8] Dewey is principally interested in the relation between the individual and society, and in the constitution of the public. He argues that there is an essential need for the improvement of methods and conditions of debate, discussion, and persuasion. Thus, the fostering of civic imaginaries is an attempt to envision and ultimately realize an inclusive and relational public by forwarding commonalities over differences.

To align with relational theorists, Dewey's prescription needs to be extended to include nonhuman actors. Associational behavior involves first and foremost the recognition that one is associated with humans and nonhumans alike, and a civic imagination is the basis of that recognition. Projects such as Growing Vine Street and the Longfellow Creek Legacy Trail in Seattle are examples of how urban water can be shaped to highlight its relational attributes. The messiness of these spaces can be off-putting and difficult to discern through conventional lenses of urban development, but they present an opportunity to teach urban residents new ways of seeing the world. They are first and foremost pedagogical tools to foster civic imagination.

A Call for Civic Expertise

Both public problem solving and democratic governance, especially when they apply to value-laden policy issues, would be better served if the technically oriented, top-down expert-client relationship were replaced by a more professionally modest but politically appropriate understanding of the expert as "specialized citizen."

—Frank Fischer[9]

The fostering of civic or ecological imaginaries is an important first step to enact a relational perspective of urban nature. In addition, there is a need for a variety of actors to turn this newfound understanding of relations into concrete projects and practices. John emphasizes the need for a "shadow community" of experts from government, business, and civil society organizations to foster collaborative political activities.[10] Although this label has a somewhat sinister air, the mission of these individuals is actually one of *humble expertise*, where specialized knowledge informs rather than dominates political deliberation. Fischer argues for a similar model of civic expertise, where specialized knowledge is oriented first and foremost toward democratic empowerment.[11] This approach follows Dewey's vision of the expert in society as that of researcher, clarifier, hypothesizer, and general assistant to resolve publicly determined problems. The public formulates a problem definition and experts then provide scenarios and possible courses of action for the public to deliberate.[12] The role of the civic expert is not to dictate political activities but rather to facilitate and guide as necessary; the civic expert is committed first and foremost to making the process successful rather than advocating for a particular outcome.[13]

This arrangement results in a blurring of the boundary between experts and nonexperts, with experts understood as specialized citizens and citizens as local experts.[14] I do not argue that specialized knowledge is no longer necessary; natural and social scientists are crucial to civic politics, although their specialized knowledge constitutes only a portion of the relations that exist in a particular locale. The tendency of these individuals should not be to fall back on the certainties of natural science or the detached gaze of social analysis, but rather to work toward elucidating relations, both contested and consensual, and to speculate on the potential consequences of reworking those relations into new configurations. Latour writes, "So the sciences are going to put into the common basket their skills, their ability to provide instruments and equipment, their capacity to record and listen to the swarming of different imperceptible propositions that demand to be taken into account."[15]

Processes of negotiation and democratic deliberation are familiar in urban theory and practice, particularly in the field of communicative planning, in which discursive forms of urban development are emphasized.[16] Planning theorist John Forester summarizes this approach:

> In cities and regions, neighborhoods and towns, planners typically have to shuttle back and forth between public agency staff and privately interested parties, between neighborhood and corporate representatives, between elected officials and civil service bureaucrats. They do not just shuttle back and forth though. Trying to listen carefully and argue persuasively, they do much more. They work to encourage practical public deliberation—public listening, learning and beginning to act on innovative agreements too—as they move project and policy proposals forward to viable implementation or decisive rejection.[17]

Forester's program of urban politics is one of negotiation and mediation between different social actors in which the expert serves as a broker of information between different stakeholders and actively engages in forging compromise. Communicative planners focus on the humans in planning processes, an approach similar to a relational understanding of social ordering and agency.[18] However, Forester and other communicative planners adhere to a decidedly anthropocentric forms of communicative rationality, an approach most closely associated with Jürgen Habermas's theory of communicative rationality.[19] Dryzek notes that for Habermas, the only voices that matter are human.[20] The relational insistence on a "more-than-human" world has significant implications for communicative rationality because it requires us to involve not only humans but also nonhumans in political deliberation.[21]

Dryzek makes an explicit link between deliberative democracy and communication with nonhumans by forwarding the notion of ecological democracy. He notes that like communicative rationality, democratic theory has also been limited to an anthropocentric focus. Rather than adopting an ecocentric perspective, as many environmentalists have done, Dryzek argues for the extension of Habermas's communicative rationality to nonhuman actors:

> The key here is to downplay "centrism" of any kind, and focus instead on the kinds of interactions that might occur across the boundaries between humanity and nature. In this spirit, the search for green democracy can indeed involve looking for less anthropocentric political forms. For democracy can exist not only among humans, but also in human dealings with the natural world—though not *in* that natural world, or in any simple *model* which nature provides for humanity. So the

key here is seeking more egalitarian interchange at the human/natural boundary; an interchange that involves progressively less in the way of human autism. In short, ecological democratization here is a matter of more effective integration of political and ecological communication.[22]

Similar to relational theorists, Dryzek argues for the rejection of essences and an embrace of the hybrid world, although he begins with the notion of deliberative democracy rather than a questioning of conventional ontological categories.

Dryzek furthers this perspective by arguing for a communicative rationality that listens to nonhuman voices. Communicating with nonhuman actors might seem to be a perverse activity, given the limitations of these actors to engage in conventional forms of communication such as speaking, reading, writing, and emotion, but Dryzek recognizes communication from nature on a large scale through climate change, desertification, deforestation, and species extinction. This form of ecological communication is evident in local hydrologic flows through processes of flooding, bank erosion, pollution concentrations, and so on. Dryzek writes:

Recognition of agency in nature therefore means that we should listen to signals emanating from the natural world with the same sort of respect we accord communication emanating from human subjects, and as requiring equally careful interpretation. In other words, our relation to the natural world should not be one of instrumental intervention and observation of results oriented to control. Thus communicative interaction with the natural world can and should be an eminently rational affair.[23]

In this view, the value of civic experts is not in the specialized knowledge that they bring to the negotiating table but rather their ability to mediate human/nonhuman conflicts through various listening processes. The municipal staff members in Austin and Seattle are the principal "listeners" of urban water flows, tasked with translating this information from the waterways, endangered species, vegetation, and so on to venues of political deliberation. Such experts can be understood to be, as Winston Churchill famously stated, "on tap, rather than on top."[24] But informal experts also play a crucial role in translating the signals of their local surroundings for political deliberation. Both can be considered civic experts tasked with mediating communication between humans and nonhumans.

A Call for Civic Experimentation

Regulations are, and always should be, in a state of flux and adjustment—on the one hand with a view to preventing newly discovered abuses, and on the other

hand with a view to opening a wider opportunity of individual discretion at points where the law is found to be unwisely restrictive.
—Frederick Law Olmsted Jr.[25]

Beyond the entrenched modes of thinking about human/nonhuman relations as well as the social structures that reify these modern dichotomies, one of the most formidable barriers to realizing new relations between humans and nature is the material obduracy of the built environment. The perennial and seemingly permanent infrastructure of the nineteenth century slowly crumbles under the feet of urban residents, but the logic embedded in these networks persists in physical form and can be reoriented only at high cost.[26] As such, there is a tendency in urban development processes to perpetuate the nature/culture relations that were instituted over a century ago simply because of the physical resistance to radical change.

The obduracy of urban infrastructure and form does not end with its materiality. One of the lessons from the Promethean Project of controlling urban water flows is that technomanagerial culture is equally resistant to change. The gradual adoption of source control strategies for urban runoff highlights the challenges associated with changing scientific and technological strategies to reflect new social understandings of human/ nature relations. Managerial obduracy includes technical experts who monopolize the design, construction, and maintenance of these networks as well as codes and regulations that embed this logic into the DNA of urban government activity. Ecological designers Sim Van der Ryn and Stuart Cowan lament the codification of urban environmental management, noting: "City planners, engineers, and other design professionals have gotten trapped in standardized solutions that require enormous expenditures of energy and resources to implement. These standard templates, available as off-the-shelf recipes, are unconsciously adopted and replicated on a vast scale. The result might be called *dumb design*: design that fails to consider the health of human communities or of ecosystems, let alone the prerequisites of creating an actual place."[27]

To counter this problem of "dumb design," perhaps better understood as a compartmentalized, topographic perspective of human/nonhuman relations, Moore argues that we need to employ experimental thinking that does not stick to the same scientific assumptions, traditional values, and social habits that created the problems in the first place.[28] This is not to advocate for the abandonment of codes and regulations altogether but rather the need for flexibility in code-making and enforcement activities to allow for customized solutions to particular situations—solutions that are frequently prohibited in existing regulatory frameworks. Technomanagerial bureaucracies tend to be overly rigid in their pursuit of certainty and prevent new relations from being tested.

This lack of experimental thinking is particularly evident in Austin, where traditional engineering practices and land use regulations dominate the urban stormwater discourse. Many of the respondents expressed a desire to try different approaches that are not allowed by existing regulations. A Watershed Department staff member summarizes this position: "Personally, I'm much more willing to let a lot of this innovative stuff go through with some level of rigor and evaluation but let's just find out, do some experimentation. It can be a big experiment about what works and what doesn't work. Right now, a huge percentage of our runoff is not treated anyway so why not do a little experimentation and then see where we are."[29] This perspective follows on Dewey's call for experimental thinking in governance: "The formation of states must be an experimental process . . . the State must always be rediscovered."[30] This process of discovery can be fostered and encouraged through civic politics that develop experimental projects of relation building. Such an approach fosters imaginative and innovative solutions that can stimulate changes in social habits and values.[31] The emergence of the Natural Drainage System Approach in Seattle—particularly its largest iteration at the High Point Redevelopment—points to a process of changing codes by experimentation from within the municipal government. Reflecting on the importance of SEA Street to municipal code building, environmental planner Eran Ben-Joseph writes, "Ultimately the endorsement and backing of local jurisdictions of unconventional practices together with their economic benefits will create momentum for change. Experimentation and fluidity would replace exacting rules. Principles would transcend codes and standards."[32]

However, the municipal experiments conducted by the City of Seattle are constrained to some extent by the existing structures of expertise and regulation. More radical experimentation is also taking place in Austin and Seattle but in another arena: the university. At the University of Washington, landscape architect Daniel Winterbottom initiated a design/build program in 1998 aimed at engaging students in collaborative community projects, similar to Spirn's work at Mill Creek.[33] Winterbottom's students work with community members to design and construct public facilities in parks and community gardens. Water flows are often a central element of these projects; rainwater harvesting systems and artistic expressions of cultural water symbols (e.g., salmon, raindrops) are used to make an explicit connection between community building and urban hydrologic flows. A University of Washington faculty member summarizes this approach: "The university setting affords the perfect opportunity to operate in left field under the guise of exploratory education. We can do laboratory work that government agencies really can't do without a whole lot of paperwork and jurisdiction. We can sort of work at the fringe but gain notoriety and credibility through the agencies."[34] Such an approach provides pilot projects for reorienting public spaces while engaging the public in infrastructure building. Likewise, the University of Texas at Austin's School of Architecture also has a design/build program; although it does not focus specifically on landscape and water flows, stormwater issues are often an integral part of the designs.[35]

Experimentation here does not involve the isolated laboratory work of natural scientists but rather a certain logic of practice that is recursive and iterative—an art of experimental thinking.[36] Geographer Steve Hinchliffe sums up this experimental approach, stating, "The injunction must be to join the doings, to experiment, to engage in the doings of environments, to environ them in different and better ways."[37] In this sense, civic politics is less about bureaucratic procedures and structures and more about experimentation, openness to unanticipated outcomes, and acceptance of uncertainty in decision-making processes.[38]

Conclusions

What is often being argued, it seems to me, in the idea of nature is the idea of man; and this not only generally, or in ultimate ways, but the idea of man in society, indeed the ideas of kinds of societies.
—Raymond Williams[39]

In 2004, the City of Austin opened a new city hall in the central core of the city within walking distance of Barton Springs. The design of the building reflects the porous and fractured geologic layers of the Edwards Aquifer. Likewise, Seattle's new city hall, inaugurated in 2003, is tied to the neighboring Seattle Justice Center and a civic plaza by an artificial waterway that wends its way through the buildings. The waterway serves as a figurative celebration of water flowing from the Cascade Mountains to Puget Sound.[40] These are aesthetic architectural features, to be sure, but they serve to embed water flows in the structure of these civic buildings (see figure 8.1). Moreover, they are a deliberate and visible recognition of the importance of nature to the political culture of these cities. They reflect the intertwining of governance, technology, landscape, and

Figure 8.1
The design of the new Austin City Hall reflects the layers of the Edwards Aquifer (top); an artificial waterway runs through Seattle City Hall (bottom).

community as highlighted by urban runoff activities in Austin and Seattle. From one perspective, this connectedness can be seen as detrimental—the root of wicked environmental problems that confound technomanagerial attempts to devise amenable solutions. Conversely, this connectedness can be leveraged as an opportunity to redefine communities by recognizing that nature is an indelible part of urban life that needs to be acknowledged and nurtured.

Creating new forms of urban nature may seem like a heroic endeavor, but civic politics are frequently catalyzed by ordinary citizens who imagine new hybrid relations, identify humble experts and sympathetic neighbors to assist them in design activities, and ultimately engage in small but important experiments to test their ideas. Those interested in enacting a relational politics can leverage opportunities in local community groups such as neighborhood associations in which they can initiate projects in urban waterways, community gardens, pocket parks, abandoned industrial sites, and other interstitial urban spaces. Activities might begin with an environmental cleanup and gradually move toward constructive restoration activities, or they might begin with a public art project or a community garden that feeds into other civic projects. The emphasis of these activities is not on the final outcome, as is common in conventional politics, but rather on the practice of continuous engagement with our human and nonhuman neighbors. The realization of more sustainable urban futures requires a new form of politics, one that can engage with both the human and nonhuman simultaneously. Such a political transition will not be speedy—indeed, it will likely be measured in generations or even centuries—but it is through these small, local actions that we will begin to recognize and rework the relations that bind the city into a whole, and through these activities, find our place in the world.

Notes

Preface

1. Guillerme 1994, 7.

2. Key texts include Spirn 1984, 1998; Cronon 1991; Hough 1995; Thompson and Steiner 1997; Braun and Castree 1998; Castree and Braun 2001; Guy, Marvin and Moss 2001; Moore 2001; Gandy 2002; Johnson and Hill 2002; Whatmore 2002; Desfor and Keil 2004; Latour 2004; Swyngedouw 2004; Brand 2005; Kaika 2005; Heynen, Kaika, and Swyngedouw 2006; Murdoch 2006; Moore 2007; and Hinchliffe 2007.

3. Swyngedouw and Kaika 2000, 568.

4. See Dryzek 1990, 2000; John 1994; Mazmanian and Kraft 1999; Fischer 2000; Shutkin 2000; Barber 2003; Durant, Fiorino, and O'Leary 2004; Agyeman 2005; Dobson and Bell 2006; Cannavò 2007; and Kütting and Lipschutz 2009.

5. Since 1991, the U.S. Environmental Protection Agency has recognized outstanding municipal stormwater programs and projects through its National Storm Water Management Excellence Program (see U.S. EPA 2008). Likewise, the Natural Resource Defense Council has published two reports of stormwater best practices and identified several municipal, county, and state government agencies that are attempting to address urban water quality issues. The municipalities of Austin and Seattle are identified by each of these organizations as promoting and implementing innovative approaches to stormwater management. See Lehner et al. 1999 and Kloss and Calarusse 2006.

6. One of the most well-publicized rankings of U.S. sustainable cities is the SustainLane US City Rankings, in which both Austin and Seattle are identified as leaders in creating more sustainable forms of living. See SustainLane 2008. These rankings are based on a variety of metrics (environmental quality, planning and land use, affordability, transportation, green building, knowledge base, etc.) and are informed to some degree by urban runoff activities, particularly with respect to land use, water quality, and municipal governance.

7. On sociotechnical urban research methodologies, see Guy and Karvonen 2011.

8. Karvonen 2010.

9. Swyngedouw 2007, 24.

1   The Dilemma of Water in the City

1. Mumford 1956, 386.

2. Gandy 2002.

3. White 1996, 121. For a compelling argument about the relationship between societal progress and the control of nature, see Marx 1996.

4. Nye 2003b, 10.

5. Boyer 1983. On large technical systems, see Hughes 1987; Mayntz and Hughes 1988; Summerton 1994; Coutard 1999; Jacobson 2000; Graham and Marvin 2001; Coutard, Hanley, and Zimmerman 2005; and Hommels 2005.

6. Schultz and McShane 1978.

7. Schultz and McShane 1978; Tarr et al. 1984; Tarr 1988.

8. Dryzek 1997.

9. Hughes 1987.

10. Burian et al. 1999. For an example, see Jones 1967.

11. Kaika 2005. The term "Promethean" refers to Prometheus, a character from Greek mythology who was punished by the gods for stealing fire from the heavens and giving it to humans. The subtitle of Mary Shelley's 1818 novel *Frankenstein* was "The Modern Prometheus," reflecting the hubris of human attempts to control nature. Dryzek (1997) characterizes the Promethean view of the world as a cornucopia of abundance with nature having adequate negative feedback mechanisms to correct for human abuses.

12. This third stage of the Promethean Project is analogous to sociologist Ulrich Beck's Risk Society thesis, particularly the argument that scientific and technological development created unintended side effects that threaten human existence. See Beck 1992, 1999.

13. Swyngedouw 2004; Kaika 2005.

14. Kaika 2006, 276.

15. For an overview of engineering history, see Kirby and Laurson 1932; Kirby 1956; and Reynolds 1991.

16. Quoted in Wurth 1996, 129.

17. The statement also provides fodder for the ecofeminist critique of a masculine labor engaged in the submission of a feminine Mother Nature. See Merchant 2003.

18. Kaika 2006, 276.

19. Melosi 1996.

20. Schultz and McShane 1978, 399 and 411.

21. Gandy 2004, 373.

22. Melosi 2000.

23. See Walesh 1989; Burian et al. 1999; Burian et al. 2000; Burian and Edwards 2002; Novotny 2003; and Butler and Davies 2004.

24. Tarr 1979; Burian et al. 1999; Melosi 2000.

25. Melosi 2000.

26. Tarr et al. 1984.

27. Melosi 2000.

28. Tarr and Konvitz 1987.

29. Tarr 1979.

30. Seattle is an example of a city with three distinct drainage networks, as described in chapter 6.

31. Jones and Macdonald 2007.

32. Andoh 2002. Also see Chatzis 1999.

33. For recent scholarship on the Los Angeles River, see Desfor and Keil 2004; Gandy 2006a; Gottlieb 2007; and Wolch 2007.

34. Gandy 2006a, 141.

35. Melosi 2000.

36. Tarr 1979.

37. Melosi 2000.

38. Burian et al. 1999.

39. Melosi 2000.

40. Novotny 2003.

41. Weibel, Anderson, and Woodward 1964.

42. U.S. Congress 1972.

43. Hansen 1994.

44. U.S. EPA 1983.

45. Willis 2006.

46. Willis 2006.

47. Melosi 2000, 292–293.

48. Melosi 1996, 2000.

49. Dryzek 1997, 45.

50. Girling and Kellett 2005. On the notion of the hydrologic cycle, see Linton 2008.

51. McShane 1979.

52. A primary example of the tight relationship between sewers and roads was the shift in executive control of the District of Columbia in the 1870s to the Army Corps of Engineers, who built a sewer system as well as the first successful asphalt pavements in the United States. See McShane 1979. On the history of asphalt paving, see McShane 1994 and Holley 2003.

53. Ben-Joseph 2005. Dolores Hayden (2004) notes that a consortium of lobbying groups—including auto and truck manufacturers, tire makers, gas and oil interests, highway engineers, and paving contractors—influenced the Federal Highway Administration to expand its road-building activities in the 1940s and 1950s, which has resulted in almost four million linear miles of public roads crisscrossing the United States, including the Interstate Highway System, with 94 percent of those miles paved with asphalt. Landscape architect Bruce Ferguson (2005) estimates that the paved area of the United States grows by 250 square miles per year.

54. Shuster et al. 2005. The emphasis on impervious surface coverage has generated a great deal of controversy for urban drainage professionals. Environmental engineer Vladimir Novotny (2003) refers to it as a "simple villain" because it does not take into account the complexities of urban drainage issues and regulation of impervious cover limits can lead to suburban sprawl. For various perspectives on impervious cover, see Arnold and Gibbons 1996; Schueler and Claytor 1997; Brabec, Shulte, and Richards 2002; Coffman 2002; Novotny 2003; Duany and Brain 2005; and Shuster et al. 2005.

55. Ferguson 2002; Novotny 2003.

56. Hough 1995, 47.

57. Apfelbaum 2005.

58. Sansalone and Hird 2003. For various cost estimates of stormwater management, see Girling and Kellett 2002; Heaney, Sample, and Wright 2002; and Weiss, Gulliver, and Erickson 2007.

59. Brabec, Shulte, and Richards 2002.

60. Baker 2007.

61. The most extreme example of the Promethean Project as it relates to stormwater management is the construction of deep storage tunnels in cities such as Chicago, Milwaukee, and Portland, Oregon. These enormous structures capture millions of gallons of stormwater from combined sewer overflows but are incredibly expensive to construct. See Kloss and Calarusse 2006.

62. Novotny 2003.

63. See Hughes 1987 and Moss 2000.

64. Hansen 1994; Baker 2007.

65. Apfelbaum 2005, 323.

2   Urban Runoff and the City of Relations

1. Spirn 1984, 239.

2. Hansen 1994.

3. U.S. EPA 2000; Andoh 2002; Butler and Davies 2004.

4. Kloss and Calarusse 2006.

5. Coffman 2002, 97.

6. Hurley and Stromberg 2008. The field of ecological engineering is often traced back to H. T. Odum, brother of famed ecologist Eugene Odum. See Mitsch and Jørgensen 1989; Schulze 1996; and Kangas 2004.

7. Dryzek 1997; Moss 2000.

8. For a recent biography of Olmsted, see Rybczynski 1999.

9. Boyer 1983.

10. Quoted in Howett 1998, 81.

11. Spirn 1996.

12. Meyer 1994.

13. Spirn 1996, 111.

14. Howett 1998, 84.

15. Howett 1998.

16. Schultz 1989, 155.

17. Howett 1998.

18. McHarg 1969.

19. For contemporary examples of ecological planning practitioners that have adopted McHarg's natural science approach, see Collins et al. 2000; Grimm et al. 2000; Pickett et al. 2001; and Alberti et al. 2003.

20. McHarg 1969, 165.

21. Howett 1998.

22. Forsyth 2005; Steiner 2005.

23. The Woodlands and a number of other ecological planning projects by McHarg have much in common with advocates of source control approaches to stormwater management as described previously. However, the latter hardly ever cite McHarg as an influence due to disciplinary biases.

24. Forsyth 2005; Steiner 2005.

25. Duany and Brain 2005, 296. The Woodlands development was introduced to a new generation as the unnamed subject of rock band Arcade Fire's 2010 album *The Suburbs*.

26. See McHarg 2006 and Forsyth 2005. Ecological planner Frederick Steiner, one of McHarg's most famous students, argues that "while [McHarg's] speeches and writings were full of bold rhetoric, his projects tended to be more down-to-earth and practical" (Steiner, email message to author, May 7, 2008).

27. Corner 1997; Howett 1998.

28. Mozingo 1997, 46.

29. Spirn 1998, 23.

30. Bartlett 2005, 1.

31. See Daily 1997.

32. Murdoch 2006.

33. Whatmore 1999.

34. Meyer 1997, 46.

35. Keil and Graham 1998, 102.

36. Spirn 1984, 5.

37. Braun 2005a, 640.

38. See Braun and Castree 1998; Latour 1998, 2004; Castree and Braun 2001; Whatmore 2002; and Murdoch 2006.

39. Braun and Wainwright 2001.

40. Braun 2005b, 835.

41. Haraway 1994.

42. Whatmore 2002.

43. Cronon 1995, 90.

44. Murdoch 2001.

45. Latour 2004, 232.

46. In human and cultural geography, see Harvey 1996; Graham and Healey 1999; Whatmore 2002; Murdoch 2006; Latour 2004; Harrison 2007; Thrift 2007; M. Jones 2009; and O. Jones 2009. In landscape theory, see Spirn 1984, 1998; Meyer 1994, 2005; Hough 1995; Mozingo 1997; Ndubisi 1997; and Howett 1998.

47. Quoted in Worster 1990, 15.

48. Latour 1998.

49. Environmental writer Barry Commoner also promoted a relational notion of ecology beginning in the 1970s when he famously stated, "Everything is connected to everything else" (quoted in Hough 1995).

50. Quoted in Steiner 2002, 1.

51. Ndubisi 1997, 10–11; emphasis in original.

52. Bakker and Bridge 2006, 16.

53. Note the overlap with deconstructivists such as Gilles Deleuze and Félix Guattari (see Deleuze and Guattari 1987). The term "hybrid" is itself a problematic term because it presupposes purified categories that are combined or mixed to form an unpurified world. Rather than adopting an entirely new vocabulary to understand these ideas, I acknowledge the inherent deficiencies in conventional language but continue to use these problematic terms to create a more readable account of urban nature.

54. Meyer 1994, 21.

55. Hough 1995, 30.

56. Murdoch 1998, 358.

57. Murdoch 1998. On Euclidean space, see Meyer 1994; Graham and Healey 1999; Law 1999; Bingham and Thrift 2000; Law and Urry 2004; Murdoch 2006; and M. Jones 2009.

58. Latour 1993, 118.

59. Meyer 1994, 14.

60. I will address some of the tensions between New Urbanism and stormwater management in chapter 6.

61. Meyer 1994, 15.

62. Murdoch 2006.

63. Law 1999.

64. Murdoch 2006.

65. Whatmore 1999.

66. Haraway 1988.

67. Lovelock 1979, 2006.

68. Thrift 1999.

69. Murdoch 1997.

70. Howard [1898] 1965; Corbett and Corbett 2000; Francis 2003.

71. Lenz 1990.

72. Francis 2003.

73. Corbett and Corbett 2000.

74. For the postoccupancy study of Village Homes, see Lenz 1990. For a New Urbanist critique of urban ecology, see Duany and Brain 2005.

75. In 1999, the Corbetts were honored as "Heroes of the Planet" by *Time* magazine.

76. Corbett and Corbett 2000. Mike Corbett served as city council member and mayor of Davis in the late 1980s.

77. Thayer 1994. Robert Thayer, a noted landscape theorist, is a long-time resident of Village Homes and has promoted the project as an exemplary sustainable development example for almost three decades.

78. Francis 2003.

79. Francis 2003.

80. See Dooling 2008.

81. See Corbett and Corbett 2000; Gause 2003; and Pieranunzi 2006.

82. For examples of the contemporary landscape architecture discourse, see Thompson and Steiner 1997; Johnson and Hill 2002; and Waldheim 2006.

83. For examples of contemporary landscape architecture approaches to stormwater management, see France 2005.

3   Saving the Springs

1. U.S. Census 2000.

2. Hart 1974, 24.

3. The original street grid of Austin was a lattice of nature, with north–south streets named after Texas rivers (e.g., Colorado, Guadalupe, San Jacinto) and

east–west streets named after regional tree species (e.g., Pecan, Cedar, Hickory). See Swearingen 2010. The latter streets were converted to ordinal names in the late nineteenth century.

4. Quoted in Horizons '76 Committee (1976, 14) and in Pipkin and Frech 1993.

5. Quoted in Horizons '76 Committee (1976, 14).

6. City of Austin 1992.

7. City of Austin 1992, 11.

8. Bartowski and Swearingen 1997.

9. Quoted in Pipkin and Frech 1993, 69. Humphrey (1985) notes that the pool was off limits to African Americans as late as 1959, so the widely held notion that the springs has always been an egalitarian, democratic locale is inaccurate.

10. Quoted in Pipkin and Frech 1993, 105.

11. Swearingen 1997.

12. Aus71.

13. See Humphrey 1985; Kanter 2004; Wassenich 2007; and Long 2010.

14. Clark-Madison 2002, 22.

15. Aus62.

16. City of Austin 1992, 5.

17. One of the most distinctive elements of the pool is its natural bottom, which swimmers can feel under their feet.

18. On the history of the greenbelt, see Swearingen 2010.

19. On the notion of "green romanticism," see Dryzek 1997.

20. Banks and Babcock 1988.

21. Hadden and Cope 1983.

22. Jones 1982, 297.

23. Sevcik 1992.

24. Orum 1987; Humphrey 2001.

25. Humphrey 2001.

26. Just downstream of the Highland Lakes, Lady Bird Lake (formerly Town Lake) abuts Austin's downtown. The lake was created in 1960 with the completion of the Longhorn Dam but is not an official part of the Highland Lakes chain. The lake is named after Lyndon Johnson's wife because of her tireless efforts to preserve nature in Austin as well as the United States. See Swearingen 2010.

27. Orum 1987, 229.

28. Quoted in Pipkin 1995, 35.

29. Orum 1987; Swearingen 2010.

30. Quoted in Banks and Babcock 1988, 178–179.

31. Humphrey 2001.

32. Smith 2001.

33. Oden 1997.

34. See Logan and Molotch 1987; Jonas and Wilson 1999; and Swearingen 2010.

35. For a contemporary perspective on the U.S. conservation/preservation debate, see Cannavò 2007.

36. Orum 1987.

37. Humphrey 2001.

38. Orum 1987, xiii.

39. Swearingen 1997, 2010.

40. Sale 1985; Dryzek 1997.

41. Quoted in City of Austin 1992, 28.

42. Throughout this study, I focus specifically on the portion of the Hill Country that is located directly to the west of Austin, specifically the western part of Travis County and eastern Hays County. The actual boundaries of the region are a continual subject of debate.

43. Doughty 1983.

44. Caro 1982, 11.

45. Doughty 1983.

46. For a photographic essay on the transformation of the U.S. landscape from rural to suburban, see MacLean 2003.

47. Aus75.

48. See Haraway 1991.

49. Ross and Sih 1997. The Edwards Aquifer serves as a local analogy to the global issue of climate change and the challenge of understanding a complex and dynamic geophysical system.

50. City of Austin 1992.

51. A frequently overlooked point is that the Barton Springs section of the Edwards Aquifer is the sole source of water supply to more than 50,000 residents in the region. See Price 2006b. This population is small when compared with the 800,000 residents who rely on the City of Austin water supply sourced from the Colorado River upstream of Barton Creek. See Austin Water Utility 2007. Conversely, the water supply for San Antonio is dependent on the Southern section of the Edwards Aquifer, a relatively independent hydrologic system.

52. The local dialog on the negative impacts of urban growth did not begin in the 1970s and can be traced back to the 1880s when Austin had only 14,000 residents. See Hart 1972.

53. Shrake 1974, 26.

54. City of Austin 1980.

55. Hurn 1983; Butler and Myers 1984; Swearingen 2010.

56. Aus75.

57. Oden 1997; Swearingen 1997; Swearingen 2010.

58. Butler and Myers 1984.

59. Cities in Texas have the ability to deny a MUD proposal, but if denied, they are legally obligated to provide infrastructure services. So it is often not a question of whether to build infrastructure but who will provide it. See Butler and Myers 1984.

60. Quoted in Shevory 2007.

61. Moore 2007, 48.

62. See Guy, Graham, and Marvin 1997 and Graham and Marvin 2001.

63. Butler and Myers 1984.

64. Aus64.

65. Aus65.

66. Public Employees for Environmental Responsibility 2008.

67. Pincetl 2004.

68. Butler and Myers 1984.

69. Clark-Madison 2001.

70. U.S. Census 2000. These statistics are somewhat misleading because the geographic area of Texas cities tends to be much larger than other U.S. cities and more comparable to metropolitan areas (a separate census unit).

71. Clark-Madison 2001.

72. Aus62.

73. City of Austin 2001a.

74. Steiner 2006. This is the same firm that designed the Woodlands development, as described in chapter 2. Many people call this the "McHarg Plan" because of his central involvement in its creation.

75. Marsh and Marsh 1995.

76. City of Austin 2001a.

77. Swearingen 2010.

78. Moore 2007. On the cultural implications of development codes, see Ben-Joseph 2005.

79. Moore 2007.

80. The name of the organization is a persuasive framing of the topic, suggesting an imminent threat to a shared resource.

81. Aus71.

82. Quoted in Pipkin and Frech 1993, 86.

83. Aus65.

84. Aus60.

85. Aus71.

86. Marsh and Marsh 1995.

87. Aus65.

88. Swearingen 2010.

89. Quoted in *Austin Chronicle* 2002.

90. Jacobs 2005, 20.

91. Aus62.

92. Note that the species name includes the SOS acronym to acknowledge the work of local environmental activists.

93. Lieberknecht 2000.

94. U.S. Department of Interior 1997.

95. Lieberknecht 2000, 150–151. Lieberknecht notes that a popular T-shirt states, "The Barton Springs Salamander—not just an amphibian, an Austin original," further tying the springs to Austin's unique culture (2000, 156).

96. See Dryzek 1997 and Brulle 2000.

97. Clark-Madison 2002.

98. Quoted in Price 2007.

99. Quoted in Price 2007.

100. Aus75.

101. Hess 2007.

102. See Alexander 2006 and Selden 2006.

103. Aus66.

104. Aus66.

105. See <http://www.liveablecity.org> and <http://www.hillcountryconservancy.org>.

106. Aus62.

107. City of Austin 2007c. Note that some of the conservation land purchases are outside of the Austin city limits, an activity generally reserved for protection of municipal drinking water systems such as New York City's Croton System and the Owens Valley water supply for Los Angeles.

108. Duerksen and Snyder 2005.

109. Quoted in Katz 1998.

110. Aus71.

111. Shutkin 2000, 233.

112. For information on the regional water quality plan, see <http://www.waterqualityplan.org>.

113. Murdoch 2006.

114. Aus74.

115. Aus69.

116. City of Austin 2007a; Aus69.

117. Aus73.

118. Gregor 2007.

119. Price 2006b.

120. Swearingen 2010, 188.

4   After the Flood

1. Horizons '76 Committee 1976, 6.

2. Koch and Fowler 1928, 3.

3. See City of Austin 1961, 1980, and 1988. Two Austinites, Russell and Janet Fish, are credited with coining the term "hike-and-bike" to describe multiuse linear park facilities. They personally financed the original trail system along Shoal Creek. See Horizons '76 Committee 1976 and Swearingen 2010.

4. Beatley, Brower, and Lucy 1994.

5. Quoted in Jones 1982, 17.

6. Hamilton 1913.

7. This historic process of locating sanitary sewers in the creek beds is now being reversed by the Austin Clean Water Program. Beginning in 2001 under a federal government mandate, the goal of the program is to improve water quality of the urban waterways by replacing leaking sanitary sewer lines and relocating them outside of the natural drainage channels. To date, the $400 million program has replaced almost twenty miles of pipe in seventy neighborhoods and has significantly reduced sewage overflow volumes and incidents. See City of Austin 2010.

8. Some of the first sewer pipes from the late nineteenth and early twentieth century are still in use in the downtown and central core neighborhoods. See City of Austin 2001b.

9. Jones 1982.

10. For a summary of Austin flood events, see <http://www.ci.austin.tx.us/watershed/flood.htm>.

11. Aus72.

12. City of Austin 2001b.

13. Doughty 1983; Marsh and Marsh 1995.

14. In some cases, sanitary sewer lines in the creek bed have the unintended consequence of halting erosion by serving as a nonerodable creek bed.

15. City of Austin 2001b.

16. Reeves 2007.

17. See the Waller Creek Tunnel project website: <http://www.ci.austin.tx.us/wallercreek>.

18. Aus63.

19. Aus75.

20. A more comprehensive attempt to make space for water in built-up areas is being undertaken in the Netherlands with the national government's "Ruimte Voor

de Rivier (Room for the River)" program. The program involves drastic changes to create additional flood plains on the Rhine River to reduce the impacts of future climate change. See <http://www.ruimtevoorderivier.nl>.

21. On the notion of urban obduracy, see Hommels 2005.

22. Barrett, Quenzer, and Maidment 1998.

23. Lehner et al. 1999.

24. The 2007 budget for the monitoring program was about $1 million, which included $300,000 from the U.S. Geological Survey. See Glick 2007.

25. Aus60.

26. Aus65.

27. On the notion of technological momentum, see Hughes 1987.

28. The source control dialog existed in Austin as early as 1979 when a consulting firm produced a report on Barton Creek Watershed for the City of Austin and called for an approach that took advantage of natural drainage patterns. See City of Austin 1979.

29. Aus61.

30. Pasternak 1995.

31. Aus72.

32. City of Austin 2001b.

33. Aus61.

34. Aus61.

35. Matt Hollon, City of Austin Watershed Protection and Review Department, email message to author, December 19, 2007. The municipality began a related program in 1998 to characterize the drainage networks of Austin. The multimillion-dollar effort will delineate the stormwater system in the municipality's jurisdiction and allow for hydraulic modeling to determine future capacity and management needs. See Campman 2007.

36. Aus63.

37. Aus75.

38. City of Austin 2001b.

39. City of Austin 2001b.

40. City of Austin 2001b.

41. City of Austin 2001b, ES-61.

42. City of Austin 2001b.

43. Aus75.

44. Gleason 2002.

45. Aus61.

46. Aus75.

47. Duerksen and Snyder 2005.

48. Aus72.

49. Fischer 2006.

50. Butler and Myers 1984.

51. Moore 2007; Swearingen 2010. In 2009, the municipality initiated a new comprehensive planning process, and the plan is scheduled for completion in 2011.

52. Sheldon 2000. For more information on Smart Growth, see the U.S. EPA's Sustainable Communities Network website at <http://www.smartgrowth.org>.

53. Gearin 2004.

54. Aus71.

55. Swearingen 2010.

56. City of Austin 2007b.

57. City of Austin 2001b, ES-40.

58. Aus63.

59. Gearin 2004.

60. On the tensions between economic development, environmental protection, and social equity in the pursuit of sustainable development, see Campbell 1996.

61. Several pockets of African American neighborhoods existed outside of East Austin, such as Clarksville and Wheatsville, but these were exceptions to the general racial distribution of the city.

62. Koch and Fowler 1928.

63. Butler and Myers 1984.

64. Aus65.

65. Aus62.

66. See Melosi 2000 and Agyeman 2005.

67. Hadden 1997.

68. Hadden 1997, 80–81.

69. Clark-Madison 1999; Chusid 2006.

70. Almanza, Herrera, and Almanza 2003.

71. Clark-Madison 1999.

72. Aus67.

73. Aus66.

74. Aus63.

75. Quoted in Price 2006a.

76. Quoted in Price 2006a.

77. Aus75.

78. Aus75.

5 Metronatural™

1. Seattle Convention and Visitors Bureau 2006.
2. James 2006; Johnson 2006.
3. Raban 1993, 33.
4. On the notion of "metropolitan nature," see Gandy 2002, 2004, 2006b.
5. Klingle 2007.
6. Sale 1976, 6.
7. Schwantes 1989, 1–2.
8. Schwantes 1989, 368.
9. Edwards and Schwantes 1986, 3.
10. Edwards and Schwantes 1986; White 1995; Findlay 1997.
11. White 1995; Findlay 1997.
12. Sale 1976, 4.
13. Brambilla and Longo 1979, 1.
14. Richards 1981.
15. Brown 1986.
16. Seattle is partially in the rain shadow of the Olympic Mountains; if it were not, it would receive excessive rainfalls similar to those experienced on the Olympic Peninsula.
17. Brown 1986.
18. Church 1974.
19. For a detailed account of the Pacific Northwest rain culture, see Laskin 1997.
20. Quoted in Laskin 1997, 89.
21. Klingle 2007.
22. Fallows 2000, 22.
23. Williams 2005.
24. Klingle 2007.
25. Egan 1990. Historian John Findlay provides a persuasive critique of the frequent conflation of salmon and regional identity in the Pacific Northwest. See Findlay 1997.
26. Quoted in Taylor 1999, x.
27. Findlay 1997.
28. Taylor 1999.
29. Robbins 2001.
30. Edwards and Schwantes 1986. Note that Washington did not become a state until 1889, three decades after Oregon, and as a result was even further isolated from the United States as a nation.

31. Bunting 1997; Findlay 1997. The ties to the continent were solidified with the expansion of the Interstate highway system in the 1950s. Seattle is at the terminus of Interstate 90, which stretches eastward to Boston and the Atlantic Ocean, and is also on Interstate 5, which serves as the spine of the massive human settlement corridor on the West Coast extending from Mexico to Canada.

32. Edwards and Schwantes 1986.

33. Raban 2001, 44; Bunting 1997, 42.

34. Beaton 1914, 17.

35. Schwantes 1989.

36. For example, see the 1924 report *In the Zone of Filtered Sunshine: Why the Pacific Northwest Is Destined to Dominate the Commercial World* (Weber 1924).

37. Findlay 1997.

38. Schell and Hamer 1995.

39. Raban 2001, 41.

40. Kaiser 1987, 508.

41. Kaiser 1987; Furtwangler 1997; Klingle 2007.

42. Apostol 2006.

43. Raban 1993. This mirrors historian William Cronon's definitive account of Native American landscape alteration activities in the northeastern United States. See Cronon 1983.

44. Nye 2003a.

45. Nye 2003b, 8.

46. Lang 2003.

47. Thrush 2006, 89.

48. Beaton 1914, 64. Also see Klingle 2007. A long-standing aphorism to remember downtown street names (from south to north, in pairs) reflects the unfavorable topographic conditions: "Jesus Christ Made Seattle Under Protest." See Sale 1976. The street pairings are Jefferson/James, Cherry/Columbia, Marion/Madison, Spring/Seneca, University/Union, and Pike/Pine.

49. Findlay 1997.

50. Brown 1988.

51. Bunting 1997.

52. Journalist Fred Moody (2003) argues that the lack of ambition by Seattleites continued until the emergence of the so-called Seattle Silicon Rush that began in the 1980s as Microsoft and other companies reoriented the local economy from manufacturing to high-tech industries.

53. Bunting 1997.

54. Blackford 1980.

55. Beaton 1914; Berner 1991; Klingle 2006, 2007.

56. Cronon 1992; Klingle 2007.

57. Seney 1975.

58. Beaton 1914, 10.

59. Dorpat and McCoy 1998; Williams 2005.

60. Dorpat and McCoy 1998.

61. On the notion of the "ecological frontier," see Gandy 2002. The Cedar River watershed includes ninety-one thousand acres of municipal property that have restricted access to protect water quality. The municipality gradually acquired the property and funded the expenditure by selling the logging rights to the land. This is a curious strategy from a contemporary perspective because it is understood that mature tree cover is necessary to protect water quality. The most intensive clear-cutting activities were undertaken between 1900 and 1923, followed by extensive planting. The municipality allowed additional clear-cutting in subsequent decades; commercial logging was not curtailed until 1997. Overall, about 83 percent of the total forest cover has been logged. See Williams 2005 and Klingle 2007.

62. Thomson 1950; Sale 1976; Klingle 2007.

63. Reisner 1987; Melosi 1996, 2000.

64. Brown and Caldwell 1958.

65. Klingle 2007.

66. Lane 1995.

67. Reynolds 1991.

68. Thomson 1950, 14. Also see Klingle 2006, 2007.

69. Klingle 2007.

70. Beaton 1914, 68.

71. Klingle 2006, 2007.

72. Dimcock 1928.

73. Klingle 2006.

74. Klingle 2006, 2007.

75. Beaton 1914, 64.

76. Kirsch 1998.

77. Klingle 2006, 2007.

78. Nye 1999, 2003a; Kaika 2005.

79. Dimcock 1928.

80. Raban 1993, 30.

81. Puget Sound is the second largest estuary in the United States after Chesapeake Bay. See Jones 2007.

82. Dorpat and McCoy 1998; Vileisis 2000. Today, the total amount of impervious surface in the continental United States exceeds the total area of herbaceous wetlands. See Elvidge et al. 2004.

83. Dimcock 1928.

84. Dimcock 1928.

85. Thrush 2006. Today, only 2 percent of the original estuary banks still exist, and the Duwamish is home to some of the most altered landscapes in the greater metropolitan area. See Ith 2004. In 2001, portions of the Duwamish industrial corridor were designated as Superfund sites, and there are nascent efforts by a variety of governmental agencies—including the Port of Seattle, King County, and the U.S. Army Corps of Engineers—to clean up and restore the river. See Sato 1997 and Thrush 2007.

86. Oliver 2000, 227; Gandy 2006a.

87. Klingle 2007, 95.

88. Kaika 2005.

89. Klingle 2007.

90. Berner 1991; Klingle 2007.

91. Swyngedouw 2004, 184.

92. Klingle 2007.

93. Klingle 2006, 2007.

94. Bagley 1916, 354. Also see Klingle 2007.

95. Klingle 2007, 181–182.

96. Morgan 1951, 3.

97. Crowley 1993.

98. Klingle 2007, 106. For example, the new football stadium, Qwest Field, was built in the early 2000s on the reclaimed tideflats south of downtown and sits on 1,700 pilings that were driven fifty to seventy feet into the ground. See Magnusson 2002. Digging crews frequently uncovered bottle troves, garbage dumps, and organic soil deposits that were all unsuitable for creating a solid structural foundation.

99. Swyngedouw 2004.

100. Tubbs 1974.

101. Evans 1994.

102. Evans 1994.

103. Sea84.

104. Seattle Public Utilities 2005.

105. Kaika 2006, 297.

106. Klingle 2007.

107. Kruckeberg 1991.

108. Klingle 2007. The Hiram M. Chittenden Locks in the Ballard neighborhood of Seattle are a popular attraction for tourists and locals to watch salmon migrating up to Lake Washington and to admire the technical ability of engineers to regulate the hydrologic metabolism of the region.

109. Brown and Caldwell 1958.

110. Dorpat and McCoy 1998.

111. Klingle 2007.

112. Edmondson 1991.

113. Findlay 1997, 57.

114. Kaika 2005, 6.

115. Lehman 1986, 314. On the notion of "living laboratories," see Evans and Karvonen 2011.

116. In Seattle and communities surrounding Lake Washington, efforts to create a regional sewerage authority began as early as 1915. See Melosi 2000.

117. Lehman 1986.

118. Klingle 2007. Regional cooperation between municipalities was proposed as early as the 1870s in the United States to optimize municipal service provision through either cooperative agreements, annexation or consolidation into central cities or the creation of special governmental districts. However, these proposals were largely unsuccessful because of disagreement over responsibilities and costs, with special districts having the most success. See Tarr et al. 1984.

119. Quoted in Lane 1995, 4.

120. Edmondson 1991.

121. Lehman 1986.

122. Dorpat and McCoy 1998.

123. Dorpat and McCoy 1998.

124. Klingle 2007.

125. Klingle 2007.

126. Klingle 2007, 229.

127. Klingle 2007, 232.

128. Metro merged with King County in 1994 to become the King County Department of Metropolitan Services, although it is still referred to as Metro. See Lane 1995.

6 Reasserting the Place of Nature in Seattle's Urban Creeks

1. Sale 1976.

2. Brambilla and Longo 1979.

3. Sale 1976.

4. Herson and Bolland 1998.

5. O'Donnell 2004.

6. Sanders 2005.

7. For a brief overview of the Appropriate Technology movement, see Pursell 1993.

8. Callenbach 1975, 38.

9. Findlay 1997.

10. Sea708. Ernest Callenbach continues to publish both fiction and nonfiction related to his Ecotopian ideals, including ecology, low-consumption lifestyles, ecological restoration, and environmental economics. See <http://www.ernest callenbach.com>.

11. Sanders 2005, 8.

12. Hayden 1984.

13. Sanders 2005.

14. Steinbrueck 1962, 1973. Steinbrueck's son, Peter, was an influential member of the Seattle City Council from 1997 to 2007.

15. See Krenmayr 1973.

16. For a landscape perspective on the notions of human/nonhuman partnership and stewardship, see Schauman 1997.

17. Young 2003.

18. City of Seattle 2010.

19. City of Seattle 1994.

20. Seattle Public Utilities 2005.

21. City of Seattle 1992, 4; emphasis in original.

22. City of Seattle 1992, 64.

23. Sea714.

24. The City of Seattle embarked on a sewer separation program in the 1960s and 1970s but has largely abandoned this approach due to the high costs involved.

25. Inglis 2005.

26. Yocom 2007.

27. Seattle Public Utilities 2005.

28. See Chandler 2002 and Seattle Public Utilities 2005.

29. Seattle Public Utilities 2005.

30. Gaia Northwest and Seattle Public Utilities 1997.

31. Carkeek Watershed Community Action Project 1983; Gaia Northwest and Seattle Public Utilities 1997.

32. Piper's Creek Watershed Management Committee 1990.

33. Gaia Northwest and Seattle Public Utilities 1997.

34. Dietrich 2000.

35. In 1992, Malmgren was honored with the Governor's Environmental Excellence Individual Award, a commendation that she proudly displays in her dining room.

36. The environmental restoration work in Carkeek Park would likely be criticized by some environmental ethicists as "faking nature." For various perspectives on this debate, see Elliot 1997; Katz 2003; and Light 2003a.

37. For example, see Schell and Hamer 1995.

38. Quoted in Enlow 1999, 10.

39. Schell 1999.

40. Hajer 1995.

41. Quoted in Dietrich 2000.

42. Dietrich 2000.

43. Inglis 2005.

44. Johnson and Staeheli 2007.

45. Inglis 2005.

46. Hurley and Stromberg 2008.

47. Sea714.

48. Sea707.

49. Sea709.

50. Note that there is no consensus about upgrading streets in neighborhoods with informal drainage networks. Although some believe that the municipal government should honor its original commitment to upgrade the infrastructure in annexed areas of the city, others prefer the unfinished, rural feel of the streets. See Mills 2002.

51. Mills 2002.

52. Sea714.

53. Horner 2003, 8.

54. Inglis 2005.

55. Inglis 2005.

56. The lone holdout homeowner on SEA Street has still not embraced the new design. An SPU staff member (Sea62) notes, "She doesn't hate it now but she doesn't necessarily like it."

57. Sea707.

58. Hurley and Stromberg 2008.

59. Brown, Harkness, and Johnston 1998; Eisenstein 2001; Biscombe 2002.

60. De-Shalit 2003, 20.

61. Anderson 1978, 1.

62. Inglis 2005.

63. Sea62.

64. Inglis 2005.

65. Mills 2002.

66. During the Clinton era, the U.S. government embraced the New Urbanism approach for the Department of Housing and Urban Development's Hope VI projects, a major source of funding for the High Point Redevelopment project.

67. SHA is governed by a seven-member board of commissioners appointed by the mayor. See <http://www.seattlehousing.org>.

68. Duany, Plater-Zyberk, and Speck 2000; Hurley and Stromberg 2008.

69. Sea702.

70. Johnson and Staeheli 2007.

71. Sea62.

72. Master planning for the project was completed by Mithun Architects, Designers, and Planners; the drainage and site work was designed and coordinated by SvR Design Company.

73. Corbett and Corbett 2000; Duany and Brain 2005; Hurley and Stromberg 2008.

74. Johnson and Staeheli 2007.

75. Wiland and Bell 2006.

76. Inglis 2005.

77. Wiland and Bell 2006.

78. Wilson and Roth 2006.

79. Vogel 2006.

80. Johnson and Staeheli 2007.

81. Sea62.

82. Sea61.

83. Sea702.

84. Southworth and Ben-Joseph 2003; Ben-Joseph 2005; Shutkin 2005.

85. Sea62.

86. Hurley and Stromberg 2008.

87. Sea714.

88. Dryzek 1997; Bäckstrand 2003; Bäckstrand and Lövbrand 2006.

89. Clausen 1984; Klingle 2007.

90. Thornton Creek Watershed Management Committee 2000; Seattle Public Utilities 2007.

91. Mulady 2000a, 2000b.

92. Nabbefeld 1999.

93. Seattle Times 1999. Daylighting is the process of unearthing a buried waterway to restore its aesthetic character and ecological function.

94. Nabbefeld 1999.

95. Quoted in Mulady 2000c.

96. Harrell 2007.

97. Sea711.

98. The Nickels approach can be described as a "liberal-minimalist" form of governance that adopts a top-down authoritarian approach to resolve conflicts between economic development and environmental quality. See Moore 2007.

99. Sea708.

100. Sea708.

101. Seattle Times 2002.

102. Mulady 2003b.

103. Cannavò 2007.

104. Young 2003.

105. Young and Batesell 2003.

106. Mulady 2003a, 2003c.

107. Mulady 2003d.

108. Bush 2004; Mulady 2004.

109. Thornton Place 2011.

110. Boyer 2005.

111. Sea63.

112. Gaynor has been involved in many other urban drainage and creek daylighting projects in the city. See Schauman and Salisbury 1998.

113. Sea708.

114. Brand and Karvonen 2007; Karvonen and Brand 2009.

115. Seattle Public Utilities 2005.

116. Sea714.

117. SvR Design Company 2009.

118. Sea707.

119. Sea84.

120. Sea701.

121. Sea708.

122. See SvR 2009.

123. Grygiel 2009; Ma 2009; SvR 2009.

124. Thrush 2006, 91.

125. Sea63.

126. Quoted in Williams 2005, 67.

7   Politics of Urban Runoff

1. Gandy 2006b, 72–73.

2. On the notion of "alternative routes" or "competing pathways" to sustainable development, see Guy and Marvin 1999; Evans, Guy, and Marvin 2001; Guy and Marvin 2001; Moore 2001; Moore and Brand 2003; Guy and Moore 2005; and Moore 2007.

3. Latour 2004, 145; emphasis in original.

4. Shutkin 2000.

5. John 1994; Dryzek 1997; Shutkin 2000; Rubin 2002.

6. Shutkin 2005.

7. Fischer 1990, 2000; Bäckstrand and Lövbrand 2006.

8. Fischer 1990.

9. Fischer 2000.

10. Bäckstrand and Lövbrand 2006, 55.

11. Shutkin 2000; Agyeman and Angus 2003. The Big Ten environmental groups include the Environmental Defense Fund, Environmental Policy Institute, Friends of the Earth, Izaak Walton League, National Audubon Society, National Parks and Conservation Association, National Wildlife Federation, Natural Resources Defense Council, Sierra Club, and Wilderness Society.

12. Landy, Roberts, and Thomas 1994, 10.

13. Shutkin 2000.

14. Bäckstrand and Lövbrand 2006.

15. Landy, Roberts, and Thomas 1994.

16. Dryzek 1997, 81.

17. John 1994.

18. Shutkin 2000.

19. Roseland 2005, 196.

20. Shutkin 2000.

21. Landy, Roberts, and Thomas 1994; Shutkin 2000, 2005. The correlation between low-income and minority communities, low elevation, and degraded environmental conditions is not absolute. For example, Seattle's fourth Urban Legacy creek, Taylor Creek, is in a low-income, minority community near the south end of Lake Washington. A municipal staff member (Sea85) describes the creek as an "unknown gem in Seattle. It has some of the best water quality in the city."

22. John 1994.

23. Both Austin and Seattle have public education programs that provide "non-structural" solutions to stormwater management. They are not addressed in this study but are an important element of rational environmental politics.

24. Fischer 2000, 51.

25. Shutkin 2000.

26. See Hamilton 1943 and Knight 2005. Buell (2001) argues that there is more common ground between the ecocentric environmentalism of Muir and the anthropocentric environmentalism of Addams than is typically acknowledged.

27. John 1994, 2004.

28. Brulle 2000; Pellow 2002; Schlossberg 2007.

29. Fischer 2000.

30. Canovan 1999, 2.

31. This understanding of citizens as involved in scientific and technological knowledge generation is related to the European notion of civic science, which calls for expertise to be more transparent, accountable, and democratic. See Agyeman and Angus 2003; Bäckstrand 2003; Brand and Karvonen 2007; and Karvonen and Brand 2009.

32. John 1994.

33. Agyeman 2005, 105–106.

34. Agyeman notes that the Principles of Environmental Justice, drafted by environmental justice leaders in 1991, encompass both protest and constructive forms of action, but that the latter are generally not reflected in practice due to the disenfranchised role of these actors in politics and policy-making activities. See Agyeman 2005.

35. Gandy 2006b, 72–73.

36. John 1994, Landy, Roberts, and Thomas 1994.

37. John 1994; Cannavò 2007.

38. John 1994, 2004; U.S. EPA 1997; John and Mlay 1999; Landy, Susman, and Knopman 1999; Sirianni and Friedland 2001.

39. John 2004, 219.

40. John 1994.

41. John 1994, 32.

42. John 1994.

43. Sociologist Anthony Giddens is often cited as the most influential scholar of Third Way politics. See Giddens 1999, 2000.

44. John 1994, 2004.

45. See Hempel 1999; Shutkin 2000; Agyeman and Angus 2003; Light 2003b, 2006; Agyeman 2005; and Roseland 2006.

46. Dryzek 2000; Shutkin 2000; Rubin 2002.

47. Latour 2004, 89; emphasis in original.

48. Castree 2003, 204 and 209.

49. Murdoch 2006, 157; emphasis in original.

50. Reid and Taylor, 2003.

51. Rubin 2002, 349.

52. John 2004, 238.

53. Hempel 1999.

54. Luke 2009, 14.

55. Barber 2003.

56. Shutkin 2000, 31.

57. Dewey [1927] 1954, 213. Also see Cannavò 2007.

58. Rubin 2002, 336.

59. Hempel 1999, 48.

60. Light 2003b.

61. Agyeman 2005, 11.

62. Prugh, Costanza, and Daly 2000, 5.

63. Dryzek 1990, 2000; Beck 1992; Fischer 2000; Barber 2003; Dobson 2003; Luke 2009.

64. Dryzek 2000; Shutkin 2000.

65. Meadowcroft 2004.

66. Dryzek 1997, 86.

67. Dryzek 2000, 140.

68. Dryzek 1997, 95; emphasis in original.

69. On citizenship, see Shutkin 2000; Rubin 2002; Barber 2003; Dobson 2003; Dobson and Bell 2006; Light 2003b, 2006; Cannavò 2007; and Smith and Pang-sapa 2008.

70. Shutkin 2000, 243; King 2006, 180.

71. Dewey [1927] 1954; Dryzek 1997.

72. Light 2003b.

73. Agyeman identifies several local nongovernmental organizations in the United States that practice civic environmentalism, notably Alternatives for Community and Environment in Boston, Urban Ecology in Oakland, and the Center for Neighborhood Technology in Chicago. Of these organizations, only the Center for Neighborhood Technology has focused on urban runoff by developing a toolbox of green infrastructure to assist property owners, developers, and communities in adopting sustainable stormwater strategies (see <http://www.cnt.org>).

74. Spirn 1998, 2005. Spirn is also a noted relational theorist and student of Ian McHarg, as noted in chapter 2.

75. Spirn's West Philadelphia work is summarized on the West Philadelphia Landscape Project website: <http://web.mit.edu/4.243j/www/wplp/>.

76. Spirn 2005, 400.

77. Chuter 2008. A similar endeavor was initiated in the Oak Hill neighborhood of southwest Austin in 2004 (see <http://www.swgreenbelt.org>). And the Hill Country Conservancy is acquiring land for the Walk-for-a-Day Trail that is planned to begin at Barton Springs and follow waterways for a total of thirty-four miles in the Barton Springs segment of the Edwards Aquifer (see <http://www.hillcountryconservancy.org>).

78. Seattle Public Utilities 2007. Note that the High Point Redevelopment project described in chapter 6 is in the Longfellow Creek watershed.

79. Seattle Public Utilities 2007.

80. Yocom 2007.

81. Quoted in Biscombe 2002, 9.

82. True 2005.

83. Yocom 2007.

84. True 2005.

85. Lippard 1997, 286.

86. Sea81.

87. Quoted in Yocom 2007, 128.

88. Quoted in Yocom 2007, 124.

89. Yocom 2007.

90. The "P" in P-Patch refers "Picardo," the surname of the property owner of the first community garden in Seattle. See Diers 2004.

91. Diers 2004; City of Seattle 2007.

92. France and Fletcher 2005, 114.

93. Simpson 2004, 58.

94. Simpson 2004, 55.

95. Geise 2004, 33.

96. In his novel *Ecotopia: The Notebooks and Reports of William Weston*, Ernest Callenbach describes the daylighting of a creek on Market Street in downtown San Francisco. See Callenbach 1975.

97. Simpson 2004, 55.

98. Enlow 2003.

99. Geise 2004.

100. Meadowcroft 2004, 212.

101. John 2004.

102. Dryzek 1997; Bäckstrand 2003.

103. Landy, Roberts, and Thomas 1994.

104. Barber 2003.

105. Fischer 2000; Bäckstrand 2003.

106. Prugh, Costanza, and Daly 2000; Stilgoe, Irwin, and Jones 2006.

107. Brand and Karvonen 2007; Karvonen and Brand 2009.

108. Shutkin 2000; Bäckstrand 2003.

109. Dewey [1927] 1954, 207.

110. Graham and Healey 1999.

111. Dryzek 2000, 50.

112. Barber 2003.

113. See Sclove 2000 and Meadowcroft 2004.

114. Sclove 1995, 124.

115. Karvonen 2004; Walz 2005; Envision Central Texas 2008.

116. Maryman and Rottle 2006; Rottle and Maryman 2006; Open Space Seattle 2100 2008.

8 Toward the Relational City

1. Swyngedouw 2006, 107.
2. Williams 1972, 164.
3. Dagger 2003; Gandy 2006b; Murdoch 2006. On urban imaginaries, see Çinar and Bender 2007.
4. Hyysalo 2006. A number of environmental writers have forwarded notions of environmental or ecological *imagination* that have the potential to fuel ecological *imaginaries*. For example, see Worster 1993 and Buell 1995, 2001.
5. Murdoch 2006.
6. Dagger 2003.
7. Sea708.
8. Dewey [1927] 1954, 27.
9. Fischer 2000, 46.
10. John 1994, 2004.
11. Fischer 2000.
12. Hickman 1992.
13. John 2004.
14. Bäckstrand 2003.
15. Latour 2004, 138.
16. See Innes 1995; Healey 1997; and Forester 1999.
17. Forester 1999, 3.
18. Graham and Healey 1999; Murdoch 2006.
19. Habermas 1984.
20. Dryzek 2000.
21. The notion of a "more-than-human" world is from Whatmore 2002.
22. Dryzek 2000, 146; emphasis in original.
23. Dryzek 2000, 149.
24. Quoted in Stilgoe, Irwin, and Jones 2006, 17.
25. Olmsted 1916, 14.
26. Jacobson 2000; Hommels 2005.
27. Ryn and Cowan 1996, 10; emphasis in original.
28. Moore 2007.
29. Aus61.
30. Dewey [1927] 1954, 33–34.
31. Moore 2007.
32. Ben-Joseph 2005, 124.
33. Winterbottom 2005; Easton 2007.

34. Sea82.
35. See <http://www.thealleyflatinitiative.org> and Moore and Karvonen 2008.
36. Dewey [1927] 1954; Buchanan 1995.
37. Hinchliffe 2007, 191.
38. Rubin 2002.
39. Williams 1972, 150.
40. Cronenweth 2004.

References

Agyeman, Julian. 2005. *Sustainable Communities and the Challenge of Environmental Justice*. New York: New York University Press.

Agyeman, Julian, and Briony Angus. 2003. The Role of Civic Environmentalism in the Pursuit of Sustainable Communities. *Journal of Environmental Planning and Management* 46 (3): 345–363.

Alberti, Marina, John M. Marzluff, Eric Shulenberger, Gordon Bradley, Clare Ryan, and Craig Zumbrunnen. 2003. Integrating Humans into Ecology: Opportunities and Challenges for Studying Urban Ecosystems. *Bioscience* 53 (12): 1169–1179.

Alexander, Kate. 2006. Defeat Leaves SOS Bruised, Not Beaten, Sides Agree. *Austin American-Statesman*, May 15, <http://www.statesman.com/news/content/news/stories/local/05/15amendments.html> (accessed May 15, 2006).

Almanza, Susana, Sylvia Herrera, and Librado Almanza. 2003. *SMART Growth, Historic Zoning, and Gentrification of East Austin: Continued Relocation of Native People from Their Homeland*. Unpublished document.

Anderson, Stanford. 1978. People in the Physical Environment: The Urban Ecology of Streets. In *On Streets*, ed. Stanford Anderson, 1–11. Cambridge, Mass.: MIT Press.

Andoh, Robert Y. G. 2002. Urban Drainage and Wastewater Treatment for the 21st Century, *Global Solutions for Urban Drainage Conference Proceedings*, Portland, Ore.

Apfelbaum, Steven I. 2005. Stormwater Management: A Primer and Guidelines for Future Programming and Innovative Demonstration Projects. In *Facilitating Watershed Management: Fostering Awareness and Stewardship*, ed. Robert L. France, 321–333. Lanham, Md.: Rowman and Littlefield.

Apostol, Dean. 2006. Northwest Environmental Geography and History. In *Restoring the Pacific Northwest: The Art and Science of Ecological Restoration in Cascadia*, ed. Dean Apostol and Marcia Sinclair, 3–10. Washington, D.C.: Island Press.

Arnold, Chester L., Jr., and James C. Gibbons. 1996. Impervious Surface Coverage: The Emergence of a Key Environmental Indicator. *Journal of the American Planning Association* 62 (2): 243–258.

Austin Chronicle. 2002. The Battle for the Springs: A Chronology. *Austin Chronicle* 24 (49): 30–35.

Austin Water Utility. 2007. Austin Water Utility, <http://www.ci.austin.tx.us/water> (accessed December 3, 2007).

Bäckstrand, Karin. 2003. Civic Science for Sustainability: Reframing the Role of Experts, Policy-Makers and Citizens in Environmental Governance. *Global Environmental Politics* 3 (4): 24–41.

Bäckstrand, Karin, and Eva Lövbrand. 2006. Planting Trees to Mitigate Climate Change: Contested Discourses of Ecological Modernization, Green Governmentality and Civic Environmentalism. *Global Environmental Politics* 6 (1): 50–75.

Bagley, Clarence B. 1916. *History of Seattle from the Earliest Settlement to the Present Time*. Chicago: S. J. Clarke Publishing Company.

Baker, Lawrence A. 2007. Stormwater Pollution: Getting at the Source. *Stormwater* 8 (8): 16–42.

Bakker, Karen, and Gavin Bridge. 2006. Material Worlds? Resource Geographies and the "Matter of Nature." *Progress in Human Geography* 30 (1): 5–27.

Banks, James H., and John E. Babcock. 1988. *Corralling the Colorado: The First Fifty Years of the Lower Colorado River Authority*. Austin: Eakin Press.

Barber, Benjamin R. 2003. *Strong Democracy: Participatory Politics for a New Age*. Twentieth Anniversary Edition. Berkeley: University of California Press.

Barrett, Michael E., Ann M. Quenzer, and David R. Maidment. 1998. *Water Quality and Quantity Inputs for the Urban Creeks: Future Needs Assessment*. Austin: Center for Research in Water Resources, Bureau of Engineering Research, University of Texas at Austin.

Bartkowski, John P., and W. Scott Swearingen. 1997. God Meets Gaia in Austin, Texas: A Case Study of Environmentalism as Implicit Religion. *Review of Religious Research* 38 (4): 308–324.

Bartlett, Peggy F. 2005. Introduction. In *Urban Place: Reconnecting with the Natural World*, ed. Peggy F. Bartlett, 1–34. Cambridge, Mass.: MIT Press.

Beatley, Timothy, David J. Brower, and Wiliam H. Lucy. 1994. Representation in Comprehensive Planning: An Analysis of the Austinplan Process. *Journal of the American Planning Association. American Planning Association* 60 (2): 185–196.

Beaton, Welford. 1914. *The City that Made Itself: A Literary and Pictorial Record of the Building of Seattle*. Seattle: Terminal Publishing Company.

Beck, Ulrich. 1992. *Risk Society: Towards a New Modernity*. Newbury Park, Calif.: Sage Publications.

Beck, Ulrich. 1999. *World Risk Society*. Malden, Mass.: Polity Press.

Ben-Joseph, Eran. 2005. *The Code of the City: Standards and the Hidden Language of Place Making*. Cambridge, Mass.: MIT Press.

Berner, Richard C. 1991. *Seattle in the 20th Century*. Seattle: Charles Press.

Bingham, Nick, and Nigel Thrift. 2000. Some New Instructions for Travellers: The Geography of Bruno Latour and Michel Serres. In *Thinking Space*, ed. Mike Crang and Nigel Thrift, 281–301. New York: Routledge.

Biscombe, Jason R. 2002. "Celebrating Local Distinctiveness: A Case Study in the Design of a Seattle Neighborhood Creek Trail." Master's thesis, Landscape Architecture, State University of New York, Syracuse.

Blackford, Mansel G. 1980. Civic Groups, Political Action, and City Planning in Seattle, 1892–1915. *Pacific Historical Review* 49 (4): 557–580.

Boyer, M. Christine. 1983. *Dreaming the Rational City: The Myth of American Planning*. Cambridge, Mass.: MIT Press.

Boyer, Rick. 2005. Northgate Expansion Rumbles Back to Life. *Seattle Times*, May 19, <http://community.seattletimes.nwsource.com/archive/?date=20050519 &slug=northgate19> (accessed December 12, 2007).

Brabec, Elizabeth, Stacey Schulte, and Paul L. Richards. 2002. Impervious Surfaces and Water Quality: A Review of Current Literature and Its Implications for Watershed Planning. *Journal of Planning Literature* 16 (4): 499–514.

Brambilla, Roberto, and Gianni Longo. 1979. *What Makes Cities Livable? Learning from Seattle*. New York: Institute for Environmental Action.

Brand, Ralf. 2005. *Synchronizing Science and Technology with Human Behaviour*. London: Earthscan.

Brand, Ralf, and Andrew Karvonen. 2007. The Ecosystem of Expertise: Complimentary Knowledges for Sustainable Development. *Sustainability: Science, Practice, and Policy* 3 (1): 21–31.

Braun, Bruce. 2005a. Environmental Issues: Writing a More-Than-Human Urban Geography. *Progress in Human Geography* 29 (5): 635–650.

Braun, Bruce. 2005b. Writing Geographies of Hope. *Antipode* 37 (4): 834–841.

Braun, Bruce, and Noel Castree, eds. 1998. *Remaking Reality: Nature at the Millennium*. New York: Routledge.

Braun, Bruce, and Joel Wainwright. 2001. Nature, Poststructuralism, and Politics. In *Social Nature: Theory, Practice, and Politics*, ed. Noel Castree and Bruce Braun, 41–63. Malden, Mass.: Blackwell.

Brown, Brenda, Terry Harkness, and Doug Johnston. 1998. Eco-Revelatory Design: Nature Constructed/Nature Revealed. *Landscape Journal* 17 (2): x–xv.

Brown, Richard Maxwell. 1986. Rainfall and History: Perspectives on the Pacific Northwest. In *Experiences in a Promised Land: Essays in Pacific Northwest History*, ed. G. Thomas Edwards and Carlos A. Schwantes, 13–27. Seattle: University of Washington Press.

Brown, Richard Maxwell. 1988. The Great Raincoast of North America: Toward a New Regional History of the Pacific Northwest. In *The Changing Pacific Northwest: Interpreting Its Past*, ed. David H. Stratton and George A. Frykman, 40–53. Pullman: Washington State University Press.

Brown and Caldwell. 1958. *Metropolitan Seattle Sewerage and Drainage Survey 1956–1958: A Report for the City of Seattle, King County and the State of*

Washington on the Collection, Treatment and Disposal of Sewage and the Collection and Disposal of Storm Water in the Metropolitan Seattle Area. Seattle: City of Seattle.

Brulle, Robert. 2000. *Agency, Democracy and Nature: The U.S. Environmental Movement from a Critical Theory Perspective.* Cambridge, Mass.: MIT Press.

Buchanan, Richard. 1995. Wicked Problems in Design Thinking. In *The Idea of Design: A Design Issues Reader*, ed. Victor Margolin and Richard Buchanan, 3–20. Cambridge, Mass.: MIT Press.

Buell, Lawrence. 1995. *The Environmental Imagination: Thoreau, Nature Writing, and the Formation of American Culture.* Cambridge, Mass.: Belknap Press.

Buell, Lawrence. 2001. *Writing for an Endangered World: Literature, Culture, and Environment in the U.S. and Beyond.* Cambridge, Mass.: Belknap Press.

Bunting, Robert. 1997. *The Pacific Raincoast: Environment and Culture in an American Eden, 1778–1990.* Lawrence, Kans.: University Press of Kansas.

Burian, Steven J., and Findlay G. Edwards. 2002. Historical Perspectives of Urban Drainage. In *Urban Drainage 2002*, ed. Eric W. Strecker and Wayne C. Huber, 1–16. Portland, Ore.: American Society of Civil Engineers.

Burian, Steven J., Stephan J. Nix, S. Rocky Durrans, Robert E. Pitt, Chi-Yuan Fan, and Richard Field. 1999. Historical Development of Wet-Weather Flow Management. *Journal of Water Resources Planning and Management* 125 (1): 3–13.

Burian, Steven J., Stephan J. Nix, Robert E. Pitt, and S. Rocky Durrans. 2000. Urban Wastewater Management in the United States: Past, Present, and Future. *Journal of Urban Technology* 7 (3): 33–62.

Bush, James. 2004. Northgate South Lot Solution on Horizon. *The North Seattle Sun*, July, 1–7.

Butler, David, and John W. Davies. 2004. *Urban Drainage.* 2nd ed. New York: E & FN Spon.

Butler, Kent S., and Dowell Myers. 1984. Boomtime in Austin, Texas: Negotiated Growth Management. *Journal of the American Planning Association* 50 (4): 447–458.

Callenbach, Ernest. 1975. *Ecotopia: The Notebooks and Reports of William Weston.* Berkeley: Banyan Tree Books.

Campbell, Scott. 1996. Green Cities, Growing Cities, Just Cities?: Urban Planning and the Contradictions of Sustainable Development. *Journal of the American Planning Association* 62 (3): 296–312.

Campman, Philip. 2007. Database Design and Process Modeling for a Drainage Utility GIS. *2007 ESRI International User Conference Proceedings*, <http://proceedings.esri.com/library/userconf/proc07/papers/papers/pap_1662.pdf> (accessed November 14, 2007).

Cannavò, Peter F. 2007. *The Working Landscape: Founding, Preservation, and the Politics of Place.* Cambridge, Mass.: MIT Press.

Canovan, Margaret. 1999. Trust the People! Populism and the Two Faces of Democracy. *Political Studies* 47:2–16.

Carkeek Watershed Community Action Project. 1983. *Carkeek Watershed Master Plan Proposal.* Seattle: n.p.

Caro, Robert A. 1982. *The Years of Lyndon Johnson: The Path to Power.* New York: Knopf.

Castree, Noel. 2003. Environmental Issues: Relational Ontologies and Hybrid Politics. *Progress in Human Geography* 27 (2): 203–211.

Castree, Noel, and Bruce Braun, eds. 2001. *Social Nature: Theory, Practice and Politics.* Malden, Mass.: Blackwell Publishers.

Chandler, Robert. 2002. Collision Course: ESA, CWA, and Stormwater, *Proceedings of the 9th International Conference on Urban Drainage,* Portland, Ore.

Chatzis, Konstantinos. 1999. Designing and Operating Storm Water Drain Systems. In *The Governance of Large Technical Systems,* ed. Olivier Coutard, 73–90. New York: Routledge.

Church, P. E. 1974. Some Precipitation Characteristics of Seattle. *Weatherwise* 27 (6): 244–251.

Chusid, Jeffrey. 2006. Preservation in the Progressive City: Debating History and Gentrification in Austin. *The Next American City* 12, <http://americancity .org/magazine/article/historic-preservation-preservation-in-the-progressive-city -chusid/> (accessed November 7, 2006).

Chuter, Jackie. 2008. Personal communication, June 12, City of Austin Neighborhood Planning and Zoning Department.

Çinar, Alev, and Thomas Bender. 2007. *Urban Imaginaries: Locating the Modern City.* Minneapolis: University of Minnesota Press.

City of Austin. 1961. *The Austin Development Plan.* Austin: City of Austin Department of Planning.

City of Austin. 1979. *Barton Creek Watershed Study: Consultant's Report.* Austin: City of Austin.

City of Austin. 1980. *Austin Tomorrow Comprehensive Plan.* Austin: City of Austin Department of Planning.

City of Austin. 1988. *AustinPlan.* Austin: City of Austin Department of Planning.

City of Austin. 1992. *Hill Country Oasis: Barton Springs, Barton Creek, Edwards Aquifer.* Austin: City of Austin Parks and Recreation Department.

City of Austin. 2001a. *Watershed Ordinances: A Retrospective,* <http://www .ci.austin.tx.us/watershed/ordinances.htm> (accessed October 5, 2006).

City of Austin. 2001b. *Watershed Protection Master Plan, Phase I Watersheds Report,* COA-WPD 2001–02. Austin: City of Austin Watershed Protection Department.

City of Austin. 2006. Unpublished study on land use in the Barton Springs zone of the Edwards Aquifer. Austin: City of Austin.

City of Austin. 2007a. Demographics website, <http://www.ci.austin.tx.us/ demographics/> (accessed November 29, 2007).

City of Austin. 2007b. Smart Growth Initiative, <http://www.ci.austin.tx.us/smart growth/> (accessed November 30, 2007).

City of Austin. 2007c. Wildland Conservation Division, <http://www.ci.austin .tx.us/water/wildland> (accessed November 29, 2007).

City of Austin. 2010. *Eastside Environmental News* 1, <http://www.ci.austin .tx.us/watershed/downloads/newsletter_may10.pdf> (accessed August 10, 2010).

City of Seattle. 1992. *Mayor's Recommended Environmental Action Agenda: Environmental Stewardship in Seattle.* Seattle: City of Seattle Department of Planning.

City of Seattle. 1994. *Toward a Sustainable Seattle: A Plan for Managing Growth.* Seattle: City of Seattle Department of Planning.

City of Seattle. 2007. P-Patch Program, <http://www.seattle.gov/neighborhoods/ ppatch/> (accessed February 9, 2007).

City of Seattle. 2008. Department of Planning and Development, <http://www .seattle.gov/dpd/> (accessed February 5, 2008).

City of Seattle. 2010. Neighborhood Matching Fund Overview, <http://www .seattle.gov/neighborhoods/nmf/> (accessed July 16, 2010).

Clark-Madison, Mike. 1999. Urban on the Rocks: Neighborhood Juries Still Out on Smart Growth. *Austin Chronicle,* April 30, <http://www.austinchronicle.com/ gyrobase/Issue/story?oid=oid%3A521889> (accessed August 8, 2007).

Clark-Madison, Mike. 2001. Big City Blues. *Austin Chronicle,* January 12, <http:// www.austinchronicle.com/gyrobase/Issue/story?oid=oid%3A80153> (accessed May 15, 2008).

Clark-Madison, Mike. 2002. Did SOS Matter? Has the Movement for Clean Water and Democracy Fulfilled Its Promise? *Austin Chronicle* 21 (49): 22–25.

Clausen, Meredith. 1984. Northgate Regional Shopping Center: Paradigm from the Provinces. *Journal of the Society of Architectural Historians* 43 (2): 144–161.

Coffman, Larry S. 2002. Low-Impact Development: An Alternative Stormwater Management Technology. In *Handbook of Water Sensitive Planning and Design,* ed. Robert L. France, 97–123. New York: Lewis Publishers.

Collins, James P., Ann Kinzig, Nancy B. Grimm, William F. Fagan, Diane Hope, Jianguo Wu, and Elizabeth T. Borer. 2000. A New Urban Ecology. *American Scientist* 88:416–425.

Corbett, Judy, and Michael Corbett. 2000. *Designing Sustainable Communities: Learning from Village Homes.* Washington, D.C.: Island Press.

Corner, James. 1997. Ecology and Landscape as Agents of Creativity. In *Ecological Design and Planning,* ed. George F. Thompson and Frederick R. Steiner, 81–108. New York: John Wiley and Sons.

Coutard, Olivier, ed. 1999. *The Governance of Large Technical Systems.* New York: Routledge.

Coutard, Olivier, Richard E. Hanley, and Rae Zimmerman, eds. 2005. *Sustaining Urban Networks: The Social Diffusion of Large Technical Systems.* New York: Routledge.

Cronenweth, Charles. 2004. Sustainability with a View. *Environmental Design and Construction*, November 1, <http://www.edcmag.com/Articles/Feature_Article/76b0db719c697010VgnVCM100000f932a8c0> (accessed May 14, 2008).

Cronon, William. 1983. *Changes in the Land: Indians, Colonists, and the Ecology of New England*. New York: Hill and Wang.

Cronon, William. 1991. *Nature's Metropolis: Chicago and the Great West*. New York: W. W. Norton.

Cronon, William. 1992. A Place for Stories: Nature, History, and Narrative. *The Journal of American History* 78:1347–1376.

Cronon, William. 1995. The Trouble with Wilderness; or, Getting Back to the Wrong Nature. In *Uncommon Ground: Rethinking the Human Place in Nature*, ed. William Cronon, 69–90. New York: W. W. Norton.

Crowley, Walt. 1993. *Routes: An Interpretive History of Public Transportation in Metropolitan Seattle*. Seattle: Metro Transit.

Dagger, Richard. 2003. Stopping Sprawl for the Good of All: The Case for Civic Environmentalism. *Journal of Social Philosophy* 34 (1): 28–43.

Daily, Gretchen, ed. 1997. *Nature's Services: Societal Dependence on Natural Ecosystems*. Washington, D.C.: Island Press.

de-Shalit, Avner. 2003. Philosophy Gone Urban: Reflections on Urban Restoration. *Journal of Social Philosophy* 34 (1): 6–27.

Deleuze, Gilles, and Félix Guattari. 1987. *A Thousand Plateaus: Capitalism and Schizophrenia*. Minneapolis: University of Minnesota Press.

Desfor, Gene, and Roger Keil. 2004. *Nature and the City: Making Environmental Policy in Toronto and Los Angeles*. Tucson: University of Arizona Press.

Dewey, John. [1927] 1954. *The Public and Its Problems*. Athens, Ohio: Swallow Press.

Diers, Jim. 2004. *Neighbor Power: Building Community the Seattle Way*. Seattle: University of Washington Press.

Dietrich, William. 2000. Stream Salvation. *Seattle Times*, April 16, <http://community.seattletimes.nwsource.com/archive/?date=20000416&slug=4015776> (accessed December 12, 2007).

Dimcock, Arthur H. 1928. Preparing the Groundwork for a City: The Regrading of Seattle, Washington. *Transactions of the American Society of Civil Engineers* 92:717–734.

Dobson, Andrew. 2003. *Citizenship and the Environment*. Oxford: Oxford University Press.

Dobson, Andrew, and Derek Bell, eds. 2006. *Environmental Citizenship*. Cambridge, Mass.: MIT Press.

Dooling, Sarah. 2008. Ecological Gentrification: A Research Agenda Exploring Justice in the City. *International Journal of Urban and Regional Research* 33 (3): 621–639.

Dorpat, Paul, and Genevieve McCoy. 1998. *Building Washington: A History of Washington State Public Works*. Seattle: Tartu Publications.

Doughty, Robin W. 1983. *Wildlife and Man in Texas: Environmental Change and Conservation*. College Station: Texas A&M University Press.

Dryzek, John S. 1990. *Discursive Democracy: Politics, Policy, and Political Science*. New York: Oxford University Press.

Dryzek, John S. 1997. *The Politics of the Earth: Environmental Discourses*. New York: Oxford University Press.

Dryzek, John S. 2000. *Deliberative Democracy and Beyond: Liberals, Critics, Contestations*. New York: Oxford University Press.

Duany, Andres, and David Brain. 2005. Regulating as if Humans Matter: The Transect and Post-Suburban Planning. In *Regulating Place: Standards and the Shaping of Urban America*, ed. Eran Ben-Joseph and Terry S. Szold, 293–332. New York: Routledge.

Duany, Andres, Elizabeth Plater-Zyberk, and Jeff Speck. 2000. *Suburban Nation: The Rise of Sprawl and the Decline of the American Dream*. New York: North Point Press.

Duerksen, Christopher, and Cara Snyder. 2005. *Nature-Friendly Communities: Habitat Protection and Land Use Planning*. Washington, D.C.: Island Press.

Durant, Robert F., Daniel J. Fiorino, and Rosemary O'Leary, eds. 2004. *Environmental Governance Reconsidered: Challenges, Choices, and Opportunities*. Cambridge, Mass.: MIT Press.

Easton, Valerie. 2007. A Park for All People. *Seattle Times*, November 18, <http://seattletimes.nwsource.com/html/pacificnw11182007/2004009339_pacificplife18.html> (accessed May 25, 2008).

Edmondson, W. T. 1991. *The Uses of Ecology: Lake Washington and Beyond*. Seattle: University of Washington Press.

Edwards, G. Thomas and Carlos A. Schwantes. 1986. Discovery, Exploration, and Settlement. In *Experiences in a Promised Land: Essays in Pacific Northwest History*, ed. G. Thomas Edwards and Carlos A. Schwantes, 3–12. Seattle: University of Washington Press.

Egan, Timothy. 1990. *The Good Rain: Across Time and Terrain in the Pacific Northwest*. New York: Alfred A. Knopf.

Eisenstein, William. 2001. Ecological Design, Urban Places, and the Culture of Sustainability. *SPUR Newsletter*, June, 1–5.

Elliot, Robert. 1997. *Faking Nature: The Ethics of Environmental Restoration*. New York: Routledge.

Elvidge, Christopher D., Christina Milesi, John B. Dietz, Benjamin T. Tuttle, Paul C. Sutton, Ramakrishna Nemani, and James E. Vogelmann. 2004. U.S. Constructed Area Approaches the Size of Ohio. *Eos* 85 (24): 233.

Enlow, Clair. 1999. Seattle: Small Is Still Beautiful. *Planning* 65 (3): 4–10.

Enlow, Clair. 2003. A Watershed Moment on a Belltown Street. *Seattle Daily Journal of Commerce*, February 19, <http://www.djc.com/news/ae/11142097 .html> (accessed December 12, 2007).

Envision Central Texas (ECT). 2008. Envision Central Texas, <http://www .envisioncentraltexas.org> (accessed May 15, 2008).

Evans, James, and Andrew Karvonen. 2011. Living Laboratories for Sustainability: Exploring the Politics and Epistemology of Urban Transition. In *Cities and Low Carbon Transitions*, ed. Harriet Bulkeley, Vanesa Castán Broto, Mike Hodson, and Simon Marvin, 126–141. New York: Routledge.

Evans, Robert, Simon Guy, and Simon Marvin. 2001. Views of the City: Multiple Pathways to Sustainable Transport Futures. *Local Environment* 6 (2): 121–133.

Evans, Stephen. 1994. Draining Seattle—WPA Landslide Stabilization Projects, 1935–1941. *Washington Geology* 22 (4): 3–10.

Fallows, James. 2000. Saving Salmon, or Seattle? *Atlantic Monthly* 286 (4): 20–26.

Ferguson, Bruce K. 2002. Stormwater Management and Stormwater Restoration. In *Handbook of Water Sensitive Planning and Design*, ed. Robert L. France, 11–28. New York: Lewis Publishers.

Ferguson, Bruce K. 2005. *Porous Pavements*. New York: Taylor and Francis.

Findlay, John M. 1997. A Fishy Proposition: Regional Identity in the Pacific Northwest. In *Many Wests: Place, Culture and Regional Identity*, ed. David M. Wrubel and Michael C. Steiner, 37–70. Lawrence, Kans.: University of Kansas Press.

Fischer, Frank. 1990. *Technocracy and the Politics of Expertise*. Newbury Park, Calif.: Sage Publications.

Fischer, Frank. 2000. *Citizens, Experts, and the Environment: The Politics of Local Knowledge*. Durham, N.C.: Duke University Press.

Fischer, Frank. 2006. Environmental Expertise and Civic Ecology: Linking the University and its Metropolitan Community. In *Toward a Resilient Metropolis: The Role of State and Land Grant Universities in the 21st Century*, ed. Arthur C. Nelson, Barbara L. Allen, and David L. Trauger, 83–101. Alexandria, Va.: Metropolitan Institute at Virginia Tech.

Forester, John. 1999. *The Deliberative Practitioner: Encouraging Participatory Planning Processes*. Cambridge, Mass.: MIT Press.

Forsyth, Ann. 2005. *Reforming Suburbia: The Planned Communities of Irvine, Columbia, and the Woodlands*. Berkeley: University of California Press.

France, Robert L., ed. 2005. *Facilitating Watershed Management: Fostering Awareness and Stewardship*. Lanham, Md.: Rowman and Littlefield.

France, Robert L., and David F. Fletcher. 2005. Watermarks: Imprinting Water(shed) Awareness through Environmental Literature and Art. In *Facilitating Watershed Management: Fostering Awareness and Stewardship*, ed. Robert L. France, 103–121. Lanham, Md.: Rowman and Littlefield.

Francis, Mark. 2003. *Village Homes: A Community by Design*. Washington, D.C.: Island Press.

Furtwangler, Albert. 1997. *Answering Chief Seattle.* Seattle: University of Washington Press.

Gaia Northwest, Inc., and Seattle Public Utilities. 1997. *Piper's Creek Rehabilitation: Erosion and Sedimentation Management Program and Design Manual.* Seattle: Seattle Public Utilities.

Gandy, Matthew. 2002. *Concrete and Clay: Reworking Nature in New York City.* Cambridge, Mass.: MIT Press.

Gandy, Matthew. 2004. Rethinking Urban Metabolism: Water, Space and the Modern City. *City* 8 (3): 363–379.

Gandy, Matthew. 2006a. Riparian Anomie: Reflections on the Los Angeles River. *Landscape Research* 31 (2): 135–145.

Gandy, Matthew. 2006b. Urban Nature and the Ecological Imaginary. In *In the Nature of Cities: Urban Political Ecology and the Politics of Urban Metabolism,* ed. Nik Heynen, Maria Kaika, and Erik Swyngedouw, 63–74. New York: Routledge.

Gause, Jo Allen, ed. 2003. *Great Planned Communities.* Washington, D.C.: Urban Land Institute.

Gearin, Elizabeth. 2004. Smart Growth or Smart Growth Machine? The Smart Growth Movement and Its Implications. In *Up Against the Sprawl: Public Policy and the Making of Southern California,* ed. Jennifer Wolch, Manuel Pastor, and Peter Dreier, 279–307. Minneapolis: University of Minnesota Press.

Geise, Carolyn. 2004. Denny Regrade to Belltown: An On-Going Community Journey. In *Belltown Paradise/Making Their Own Plans,* ed. Brett Bloom and Ava Bromberg, 28–39. Chicago: WhiteWalls.

Giddens, Anthony. 1999. *The Third Way: The Renewal of Social Democracy.* Malden, Mass.: Polity Press.

Giddens, Anthony. 2000. *The Third Way and Its Critics.* Malden, Mass.: Blackwell Publishers.

Girling, Cynthia, and Ronald Kellett. 2002. Comparing Stormwater Impacts and Costs on Three Neighborhood Plan Types. *Landscape Journal* 21 (1): 100–109.

Girling, Cynthia, and Ronald Kellett. 2005. *Skinny Streets and Green Neighborhoods: Design for Environment and Community.* Washington, D.C.: Island Press.

Gleason, John. 2002. *City of Austin Wet Ponds.* Austin: City of Austin Watershed Protection and Development Review Department.

Glick, Roger. 2007. Stormwater Monitoring in Austin, Texas, presentation, National Research Council's Water Science and Technology Board Meeting, May 2, Austin, Tex.

Gottlieb, Robert. 2007. *Reinventing Los Angeles: Nature and Community in the Global City.* Cambridge, Mass.: MIT Press.

Graham, Stephen, and Patsy Healey. 1999. Relational Concepts of Space and Place: Issues for Planning Theory and Practice. *European Planning Studies* 7 (5): 623–646.

Graham, Stephen, and Simon Marvin. 2001. *Splintering Urbanism: Networked Infrastructures, Technological Mobilities and the Urban Condition*. New York: Routledge.

Gregor, Katherine. 2007. Watershed Redo: How Redevelopment Can Save the Springs . . . or Not. *Austin Chronicle* 27 (9): 32–34.

Grimm, Nancy B., J. Morgan Grove, Steward T. A. Pickett, and Charles L. Redman. 2000. Integrated Approaches to Long-Term Studies of Urban Ecological Systems. *Bioscience* 50:571–584.

Grygiel, Chris. 2009. Facing Criticism, Nickels to Run on "Record of Accomplishment," *Seattle Post-Intelligencer*, April 6, <http://www.seattlepi.com/local/404839_nickels06.html> (accessed June 19, 2009).

Guillerme, André. 1994. Water for the City. *Rassegna* 57 (1): 6–21.

Guy, Simon, Stephen Graham, and Simon Marvin. 1997. Splintering Networks: Cities and Technical Networks in 1990s Britain. *Urban Studies* 34 (2): 191–216.

Guy, Simon, and Andrew Karvonen. 2011. Using Sociotechnical Methods: Researching Human-Technology Dynamics in the City. In *Understanding Social Research: Thinking Creatively about Method*, ed. Jennifer Mason and Angela Dale, 120–133. Thousand Oaks, Calif.: Sage Publications.

Guy, Simon, and Simon Marvin. 1999. Understanding Sustainable Cities: Competing Urban Futures. *European Urban and Regional Studies* 6 (3): 268–275.

Guy, Simon, and Simon Marvin. 2001. Constructing Sustainable Urban Futures: From Models to Competing Pathways. *Impact Assessment and Project Appraisal* 19 (2): 131–139.

Guy, Simon, Simon Marvin, and Timothy Moss, eds. 2001. *Urban Infrastructure in Transition: Networks, Buildings, Plans*. Sterling, Va.: Earthscan.

Guy, Simon, and Steven A. Moore. 2005. Introduction: The Paradoxes of Sustainable Architecture. In *Sustainable Architectures: Cultures and Natures in Europe and North America*, ed. Simon Guy and Steven A. Moore, 1–12. New York: Routledge.

Habermas, Jürgen. 1984. *The Theory of Communicative Action*, trans. Thomas McCarthy. Boston: Beacon Press.

Hadden, Susan G. 1997. The East Austin Tank Farm Controversy. In *Public Policy and Community: Activism and Governance in Texas*, ed. Robert H. Wilson, 69–94. Austin: University of Texas Press.

Hadden, Susan G., and Glen Hahn Cope. 1983. Planning for Growth in Austin. *Public Affairs Comment* 30 (2): 1–8.

Hajer, Maarten A. 1995. *The Politics of Environmental Discourse: Ecological Modernization and the Policy Process*. New York: Oxford University Press.

Hamilton, Alice. 1943. *Exploring the Dangerous Trades: The Autobiography of Alice Hamilton, M.D.* Boston: Little, Brown, and Co.

Hamilton, William B. 1913. *A Social Survey of Austin*. Austin: Bulletin of the University of Texas.

Hansen, Nancy Richardson. 1994. "Social/Political Dimensions of Nonpoint Pollution Planning: A Case Study from Puget Sound." Doctoral dissertation, Natural Resources and Environment, University of Michigan, Ann Arbor.

Haraway, Donna. 1988. Situated Knowledges: The Science Question in Feminism and the Privilege of Partial Perspective. *Feminist Studies* 14 (3): 575–599.

Haraway, Donna J. 1991. *Simians, Cyborgs, and Women: The Reinvention of Nature*. New York: Routledge.

Haraway, Donna J. 1994. A Game of Cat's Cradle: Science Studies, Feminist Theory, Cultural Studies. *Configurations* 1:59–71.

Harrell, Debera Carlton. 2007. Cities Copied "Seattle Way" in Planning. *Seattle Post-Intelligencer*, July 6, <http://www.seattlepi.com/local/322686_jimdiers06.html> (accessed December 12, 2007).

Harrison, Paul. 2007. "How Shall I Say It . . . ?": Relating the Nonrelational. *Environment & Planning A* 39:590–608.

Hart, Katherine. 1972. *Waterloo Scrapbook 1971–1972*. Austin: Friends of the Austin Public Library.

Hart, Katherine. 1974. *Waterloo Scrapbook 1973–1974*. Austin: Friends of the Austin Public Library.

Harvey, David. 1996. *Justice, Nature, and the Geography of Difference*. Cambridge, Mass.: Blackwell.

Hayden, Dolores. 1984. *Redesigning the American Dream: The Future of Housing, Work, and Family Life*. New York: W. W. Norton.

Hayden, Dolores. 2004. *A Field Guide to Sprawl*. New York: W. W. Norton.

Healey, Patsy. 1997. *Collaborative Planning: Making Frameworks in Fragmented Societies*. London: Macmillan.

Heaney, James P., David Sample, and Leonard Wright. 2002. *Costs of Urban Stormwater Control*. U.S. EPA report EPA-600/R-02/021, January. Cincinnati: U.S. Environmental Protection Agency.

Hempel, Lamont C. 1999. Conceptual and Analytical Challenges in Building Sustainable Communities. In *Toward Sustainable Communities: Transition and Transformations in Environmental Policy*, ed. David A. Mazmanian and Michael E. Kraft, 43–74. Cambridge, Mass.: MIT Press.

Herson, Lawrence, and John Bolland. 1998. *The Urban Web: Politics, Policy, and Theory*. 2nd ed. Chicago: Nelson-Hall.

Hess, David J. 2007. *Alternative Pathways in Science and Industry: Activism, Innovation, and the Environment in an Era of Globalization*. Cambridge, Mass.: MIT Press.

Heynen, Nik, Maria Kaika, and Erik Swyngedouw, eds. 2006. *In the Nature of Cities: Urban Political Ecology and the Politics of Urban Metabolism*. New York: Routledge.

Hickman, Larry A. 1992. Populism and the Cult of the Expert. In *Democracy in a Technological Society*, ed. Langdon Winner, 91–103. Dordrecht, Netherlands: Kluwer Academic Publishers.

Hinchliffe, Steve. 2007. *Geographies of Nature: Societies, Environments, Ecologies*. Thousand Oaks, Calif.: Sage Publications.

Holley, I. B. 2003. Blacktop: How Asphalt Paving Came to the Urban United States. *Technology and Culture* 44:703–733.

Hommels, Anique. 2005. *Unbuilding Cities: Obduracy in Urban Socio-Technical Change*. Cambridge, Mass.: MIT Press.

Horizons '76 Committee. 1976. *Austin Creeks*. Austin: Horizons '76 Committee.

Horner, Richard R. 2003. Stormwater Runoff Flow Control Benefits of Urban Drainage System Reconstruction According to Natural Principles, In *Proceedings of the 2003 Georgia Basin/Puget Sound Research Conference*, ed. T. W. Droscher and D. A. Fraser. Olympia, Wash.: Puget Sound Action Team.

Hough, Michael. 1995. *Cities and Natural Process*. New York: Routledge.

Howard, Ebenezer. [1898] 1965. *Garden Cities of To-Morrow*. Cambridge, Mass.: MIT Press.

Howett, Catherine. 1998. Ecological Values in Twentieth-Century Landscape Design: A History and Hermeneutics. *Landscape Journal* 17 (2): 80–98.

Hughes, Thomas P. 1987. The Evolution of Large Technological Systems. In *The Social Construction of Technological Systems: New Directions in the Sociology and History of Technology*, ed. Wiebe E. Bijker, Thomas P. Hughes, and Trevor J. Pinch, 51–82. Cambridge, Mass.: MIT Press.

Humphrey, David C. 1985. *Austin: An Illustrated History*. Northridge, Calif.: Windsor Publications.

Humphrey, David C. 2001. Austin, TX (Travis County), *Handbook of Texas Online*, <http://www.tshaonline.org/handbook/online/articles/hda03> (accessed February 21, 2006).

Hurley, Stephanie, and Megan Wilson Stromberg. 2008. Residential Street Design with Watersheds in Mind. In *Handbook of Regenerative Landscape Design*, ed. Robert L. France, 287–311. New York: CRC Press.

Hurn, John Andrew. 1983. "Growth Management in Austin: Water and Wastewater Service Provision." Master's report, Public Affairs, University of Texas at Austin.

Hyysalo, Sampsa. 2006. Representations of Use and Practice-Bound Imaginaries in Automating the Safety of the Elderly. *Social Studies of Science* 36 (4): 599–626.

Inglis, Darla. 2005. Natural Drainage Systems: Leading By Example. In *Management Innovation in U.S. Public Water and Wastewater Systems*, ed. Paul Seidenstat, Michael Nadol, Dean Kaplan, and Simon Hakim, 65–75. Hoboken, N.J.: John Wiley and Sons.

Innes, Judith. 1995. Planning Theory's Emerging Paradigm. *Journal of Planning Education and Research* 14 (3): 183–189.

Ith, Ian. 2004. The Road Back: From Seattle's Superfund Sewer to Haven Once More. *Pacific Northwest: The Seattle Times Magazine*, October 3, <http://seattletimes.nwsource.com/pacificnw/2004/1003/cover.html> (accessed December 12, 2007).

Jacobs, Harvey M. 2005. Claiming the Site: Evolving Social-Legal Conceptions of Ownership and Property. In *Site Matters: Design Concepts, Histories, and Strategies*, ed. Carol J. Burns and Andrea Kahn, 19–37. New York: Routledge.

Jacobson, Charles David. 2000. *Ties that Bind: Economic and Political Dilemmas of Urban Utility Networks, 1800–1900*. Pittsburgh: University of Pittsburgh Press.

James, Andrea. 2006. Seattle's New Slogan Has Some Residents Asking: Metronatural . . . Say WA? *Seattle Post-Intelligencer*, October 21, <http://www.seattlepi.com/local/289518_metronatural21.html> (accessed December 12, 2007).

John, DeWitt. 1994. *Civic Environmentalism: Alternatives to Regulation in States and Communities*. Washington, D.C.: CQ Press.

John, DeWitt. 2004. Civic Environmentalism. In *Environmental Governance Reconsidered: Challenges, Choices, and Opportunities*, ed. Robert F. Durant, Daniel J. Fiorino, and Rosemary O'Leary, 219–254. Cambridge, Mass.: MIT Press.

John, DeWitt, and Marian Mlay. 1999. Community-Based Environmental Protection: Encouraging Civic Environmentalism. In *Better Environmental Decisions: Strategies for Governments, Businesses, and Communities*, ed. Ken Sexton, Alfred A. Marcus, William Easter, and Timothy D. Burkhardt, 353–376. Washington, D.C.: Island Press.

Johnson, Bart R., and Kristina Hill, eds. 2002. *Ecology and Design: Frameworks for Learning*. Washington, D.C.: Island Press.

Johnson, Gene. 2006. Say Wa? Metronatural, Seattle's New Tourism Slogan, Fails to Impress. *Seattle Times*, October 21, <http://community.seattletimes.nwsource.com/archive/?date=20061021&slug=metronatural21> (accessed December 12, 2007).

Johnson, Richard L., and Peg Staeheli. 2007. City of Seattle—Stormwater Low Impact Development Practices, *Proceedings of the Second Annual Low Impact Development Conference*, North Carolina State University, Wilmington, N.C.

Jonas, Andrew E. G., and David Wilson, eds. 1999. *The Urban Growth Machine: Critical Perspectives Two Decades Later*. Albany: State University of New York Press.

Jones, D. E., Jr. 1967. Urban Hydrology—A Redirection. *Civil Engineering* 37 (8): 58–62.

Jones, Grant R. 2007. *Grant Jones/Jones & Jones: ILARIS: The Puget Sound Plan*. New York: Princeton Architectural Press.

Jones, Joseph. 1982. *Life on Waller Creek: A Palaver about History as Pure and Applied Education*. Austin: AAR/Tantalus.

Jones, Martin. 2009. Phase Space: Geography, Relational Thinking, and Beyond. *Progress in Human Geography* 33 (4): 487–506.

Jones, Owain. 2009. After Nature: Entangled Worlds. In *A Companion to Environmental Geography*, ed. Noel Castree, David Demeritt, Diana Liverman, and Bruce Rhoads, 294–312. Malden, Mass.: Wiley-Blackwell.

Jones, Phil, and Neil Macdonald. 2007. Making Space for Unruly Water: Sustainable Drainage Systems and the Disciplining of Surface Runoff. *Geoforum* 38:534–544.

Kaika, Maria. 2005. *City of Flows: Modernity, Nature, and the City.* New York: Routledge.

Kaika, Maria. 2006. Dams as Symbols of Modernization: The Urbanization of Nature between Geographical Imagination and Materiality. *Annals of the Association of American Geographers* 96 (2): 276–301.

Kaiser, Rudolf. 1987. Chief Seattle's Speech(es): American Origins and European Reception. In *Recovering the Word: Essays on Native American Literature*, ed. Brian Swann and Arnold Krupat, 497–536. Berkeley: University of California Press.

Kangas, Patrick C. 2004. *Ecological Engineering: Principles and Practice.* Boca Raton: Lewis Publishers.

Kanter, Alexis. 2004. Keep Austin Weird? *Daily Texan*, September 9, <http://www.dailytexanonline.com> (accessed October 10, 2006).

Karvonen, Andrew. 2004. Envisioning Process: Point/Counterpoint. *Planning Forum* 10:69–85.

Karvonen, Andrew. 2010. Metronatural™: Inventing and Reworking Urban Nature in Seattle. *Progress in Planning* 74 (4): 153–202.

Karvonen, Andrew, and Ralf Brand. 2009. Technical Expertise, Sustainability, and the Politics of Specialized Knowledge. In *Environmental Governance: Power and Knowledge in a Local-Global World*, ed. Gabriela Kütting and Ronnie D. Lipschutz, 38–59. New York: Routledge.

Katz, Cindy. 1998. Whose Nature, Whose Culture? Private Productions of Space and the "Preservation" of Nature. In *Remaking Reality: Nature at the Millennium*, ed. Bruce Braun and Noel Castree, 46–63. New York: Routledge.

Katz, Eric. 2003. The Big Lie: Human Restoration of Nature. In *Environmental Ethics: An Anthology*, ed. Andrew Light and Holmes Rolston, 390–397. New York: Blackwell.

Keil, Roger, and John Graham. 1998. Reasserting Nature: Constructing Urban Environments after Fordism. In *Remaking Reality: Nature at the Millennium*, ed. Bruce Braun and Noel Castree, 100–125. New York: Routledge.

King, Roger J. H. 2006. Playing with Boundaries: Critical Reflections on Strategies for an Environmental Culture and the Promise of Civic Environmentalism. *Ethics Place and Environment* 9 (2): 173–186.

Kirby, Richard Shelton. 1956. *Engineering in History.* New York: McGraw-Hill.

Kirby, Richard Shelton, and Philip Gustave Laurson. 1932. *The Early Years of Modern Civil Engineering.* New Haven: Yale University Press.

Kirsch, Scott. 1998. Experiments in Progress: Edward Teller's Controversial Geographies. *Cultural Geographies* 5 (3): 267–285.

Klingle, Matthew. 2006. Changing Spaces: Nature, Property, and Power in Seattle, 1880–1945. *Journal of Urban History* 32 (2): 197–230.

Klingle, Matthew. 2007. *Emerald City: An Environmental History of Seattle*. New Haven, Conn.: Yale University Press.

Kloss, Christopher, and Crystal Calarusse. 2006. *Rooftops to Rivers: Green Strategies for Controlling Stormwater and Combined Sewer Overflows*. New York: Natural Resources Defense Council.

Knight, Louise W. 2005. *Citizen: Jane Addams and the Struggle for Democracy*. Chicago: University of Chicago Press.

Koch and Fowler. 1928. *A City Plan for Austin, Texas, prepared by Koch and Fowler, engineers, for the City Plan Commission*. Dallas: Koch and Fowler.

Krenmayr, Janice. 1973. *Foot-loose in Seattle*. Seattle: n.p.

Kruckeberg, Arthur R. 1991. *The Natural History of Puget Sound Country*. Seattle: University of Washington Press.

Kütting, Gabriela, and Ronnie D. Lipshutz, eds. 2009. *Environmental Governance: Power and Knowledge in a Local-Global World*. New York: Routledge.

Landy, Marc K., Marc J. Roberts, and Stephen R. Thomas. 1994. *The Environmental Protection Agency: Asking the Wrong Questions, From Nixon to Clinton, Expanded Edition*. New York: Oxford University Press.

Landy, Marc K., Megan M. Susman, and Debra S. Knopman. 1999. *Civic Environmentalism in Action: A Field Guide to Regional and Local Initiatives*. Washington, D.C.: Progressive Policy Institute.

Lane, Bob. 1995. *Better than Promised: An Informal History of the Municipality of Metropolitan Seattle*. Seattle: King County Department of Metropolitan Services.

Lang, William. 2003. Beavers, Firs, Salmon, and Falling Water: Pacific Northwest Regionalism and the Environment. *Oregon Historical Quarterly*. Oregon Historical Society 104 (2): 151–165.

Laskin, David. 1997. *Rains All the Time: A Connoisseur's History of Weather in the Pacific Northwest*. Seattle: Sasquatch Books.

Latour, Bruno. 1993. *We Have Never Been Modern*. Cambridge, Mass.: Harvard University Press.

Latour, Bruno. 1998. To Modernise or Ecologise? That is the Question. In *Remaking Reality: Nature at the Millennium*, ed. Bruce Braun and Noel Castree, 221–242. New York: Routledge.

Latour, Bruno. 2004. *Politics of Nature: How to Bring the Sciences into Democracy*. Cambridge, Mass.: Harvard University Press.

Law, John. 1999. After ANT: Complexity, Naming and Topology. In *Actor Network Theory and After*, ed. John Law and John Hassard, 1–14. Malden, Mass.: Blackwell Publishers.

Law, John, and John Urry. 2004. Enacting the Social. *Economy and Society* 33 (3): 390–410.

Lehman, John T. 1986. Control of Eutrophication in Lake Washington. In *Ecological Knowledge and Environmental Problem-Solving: Concepts and Case Studies*, ed. National Research Council, 301–312. Washington, D.C.: National Academy Press.

Lehner, Peter, George P. Aponte Clark, Diane M. Cameron, and Andrew G. Frank. 1999. *Stormwater Strategies: Community Responses to Runoff Pollution.* New York: Natural Resources Defense Council.

Lenz, T. 1990. "A Post-Occupancy Evaluation of Village Homes, Davis, California." Master's thesis, Technical University of Munich, Germany.

Lieberknecht, Katherine. 2000. How Everything Becomes Bigger in Texas: The Barton Springs Salamander Controversy. In *Foundations of Natural Resources Policy and Management,* ed. Tim W. Clark, Andrew R. Willard, and Christina M. Cromley, 149–172. New Haven, Conn.: Yale University Press.

Light, Andrew. 2003a. Ecological Restoration and the Culture of Nature: A Pragmatic Perspective. In *Environmental Ethics: An Anthology,* ed. Andrew Light and Holmes Rolston, 398–411. New York: Blackwell.

Light, Andrew. 2003b. Urban Ecological Citizenship. *Journal of Social Philosophy* 34 (1): 44–63.

Light, Andrew. 2006. Ecological Citizenship: The Democratic Promise of Restoration. In *The Humane Metropolis: People and Nature in the 21st-Century City,* ed. Rutherford H. Platt, 169–181. Amherst: University of Massachusetts Press.

Linton, Jamie. 2008. Is the Hydrologic Cycle Sustainable? A Historical-Geographical Critique of a Modern Concept. *Annals of the Association of American Geographers* 98 (3): 630–649.

Lippard, Lucy. 1997. *The Lure of the Local: Senses of Place in a Multicentered Society.* New York: New Press.

Logan, John R., and Harvey L. Molotch. 1987. *Urban Fortunes: The Political Economy of Place.* Berkeley: University of California Press.

Long, Joshua. 2010. *Weird City: Sense of Place and Creative Resistance in Austin, Texas.* Austin: University of Texas Press.

Lovelock, James. 1979. *Gaia: A New Look at Life on Earth.* New York: Oxford University Press.

Lovelock, James. 2006. *The Revenge of Gaia: Earth's Climate Crisis and the Fate of Humanity.* New York: Basic Books.

Luke, Timothy W. 2009. Situating Knowledges, Spatializing Communities, Sizing Contradictions: The Politics of Globality, Locality and Green Statism. In *Environmental Governance: Power and Knowledge in a Local-Global World,* ed. Gabriela Kütting and Ronnie D. Lipshutz, 13–37. New York: Routledge.

Ma, Michelle. 2009. Thornton Creek Breathes Again at Northgate. *Seattle Times,* June 19, <http://seattletimes.nwsource.com/html/localnews/2009357492_thorntoncrk19m0.html> (accessed June 19, 2009).

MacLean, Alex S. 2003. *Designs on the Land: Exploring America from the Air.* New York: Thames and Hudson.

Magnusson, Jon D. 2002. Soft Soil Makes for Tough Design. *Seattle Daily Journal of Commerce,* June 27, <http://www.djc.com/news/co/11134804.html> (accessed February 19, 2008).

Marsh, William M., and Nina L. Marsh. 1995. Hydrogeomorphic Considerations in Development Planning and Stormwater Management, Central Texas Hill Country, USA. *Environmental Management* 19 (5): 693–702.

Marx, Leo. 1996. The Domination of Nature and the Redefinition of Progress. In *Progress: Fact or Illusion?*, ed. Leo Marx and Bruce Mazlish, 201–218. Ann Arbor: University of Michigan Press.

Maryman, Brice, and Nancy Rottle. 2006. A 100-Year Plan for Open Spaces in the Emerald City. *Seattle Times*, August 13, <http://seattletimes.nwsource.com/html/opinion/2003192931_sundaybrice13.html> (accessed May 25, 2008).

Mayntz, Renate, and Thomas Hughes, eds. 1988. *The Development of Large Technical Systems*. Boulder, Colo.: Westview Press.

Mazmanian, Daniel A., and Michael E. Kraft, eds. 1999. *Toward Sustainable Communities: Transition and Transformations in Environmental Policy*. Cambridge, Mass.: MIT Press.

McHarg, Ian. 1969. *Design with Nature*. Garden City, N.J.: Doubleday/Natural History Press.

McHarg, Ian. 2006. *The Essential Ian McHarg: Writings on Design and Nature*, ed. Frederick R. Steiner. Washington, D.C.: Island Press.

McShane, Clay. 1979. Transforming the Use of Urban Space: A Look at the Revolution in Street Pavements, 1880–1924. *Journal of Urban History* 5 (3): 279–307.

McShane, Clay. 1994. *Down the Asphalt Path: The Automobile and the American City*. New York: Columbia University Press.

Meadowcroft, James. 2004. Deliberative Democracy. In *Environmental Governance Revisited: Challenges, Choices, and Opportunities*, ed. Robert F. Durant, Daniel J. Fiorino, and Rosemary O'Leary, 183–217. Cambridge, Mass.: MIT Press.

Melosi, Martin. 1996. Sanitary Engineers in American Cities: Changing Roles from the Age of Miasmas to the Age of Ecology. In *Civil Engineering History: Engineers Make History*, ed. Jerry R. Rogers, Donald Kennon, Roger T. Jaske, and Francis E. Griggs Jr., 108–122. New York: ASCE.

Melosi, Martin. 2000. *The Sanitary City: Urban Infrastructure in America from Colonial Times to the Present*. Baltimore: Johns Hopkins University Press.

Merchant, Carolyn. 2003. *Reinventing Eden: The Fate of Nature in Western Culture*. New York: Routledge.

Meyer, Elizabeth K. 1994. Landscape Architecture as Modern Other and Postmodern Ground. In *The Culture of Landscape Architecture*, ed. Harriet Edquist and Vanessa Bird, 13–34. Melbourne: Edge Publishing.

Meyer, Elizabeth K. 1997. The Expanded Field of Landscape Architecture. In *Ecological Design and Planning*, ed. George F. Thompson and Frederick R. Steiner, 45–79. New York: John Wiley and Sons.

Meyer, Elizabeth. 2005. Site Citations: The Grounds of Modern Landscape Architecture. In *Site Matters: Design Concepts, Histories, and Strategies*, ed. Carol J. Burns and Andrea Kahn, 93–129. New York: Routledge.

Mills, Melanie. 2002. "Alternative Stormwater Design Within the Public Right-of-Way, A Residential Preferences Study." Master's thesis, Landscape Architecture, University of Washington, Seattle.

Mitsch, William J., and Sven Erik Jørgensen. 1989. Introduction to Ecological Engineering. In *Ecological Engineering: An Introduction to Ecotechnology*, ed. William J. Mitsch and Sven Erik Jørgensen, 3–12. New York: John Wiley and Sons.

Moody, Fred. 2003. *Seattle and the Demons of Ambition*. New York: St. Martin's Press.

Moore, Steven A. 2001. Technology, Place, and the Nonmodern Thesis. *Journal of Architectural Education* 54 (3): 130–139.

Moore, Steven A. 2007. *Alternative Routes to the Sustainable City: Austin, Curitiba, and Frankfurt*. Lanham, Md.: Lexington Books.

Moore, Steven A., and Ralf Brand. 2003. The Banks of Frankfurt and the Sustainable City. *Journal of Architecture* 8:3–24.

Moore, Steven A., and Andrew Karvonen. 2009. Sustainable Architecture in Context: STS and Design Thinking. *Science Studies* 21 (1): 29–46.

Morgan, Murray. 1951. *Skid Road: An Informal Portrait of Seattle*. New York: Viking Press.

Moss, Timothy. 2000. Unearthing Water Flows, Uncovering Social Relations: Introducing New Waste Water Technologies in Berlin. *Journal of Urban Technology* 7 (1): 63–84.

Mozingo, Louise A. 1997. The Aesthetics of Ecological Design: Seeing Science as Culture. *Landscape Journal* 16 (1): 46–59.

Mulady, Kathy. 2000a. Is It a Creek or "Drainage Ditch" Under Northgate? *Seattle Post-Intelligencer*, April 6, <http://www.seattlepi.com/business/ngat06.shtml> (accessed December 12, 2007).

Mulady, Kathy. 2000b. Northgate Shopping Center Takes Shape, But Mall Expansion Is Stuck. *Seattle Post-Intelligencer*, March 28, <http://www.seattlepi.com/business/ngat28.shtml> (accessed December 12, 2007).

Mulady, Kathy. 2000c. Thornton Creek Citizens Group Seeks Firm Ruling. *Seattle Post-Intelligencer*, June 20, <http://www.seattlepi.com/business/crek20.shtml> (accessed December 12, 2007).

Mulady, Kathy. 2003a. City Council Discloses Own Northgate Plan. *Seattle Post-Intelligencer*, November 13, <http://www.seattlepi.com/local/148095_northgate13.html> (accessed December 12, 2007).

Mulady, Kathy. 2003b. Creek Restoration Issue Struck from Ballot. *Seattle Post-Intelligencer*, July 31, <http://www.seattlepi.com/local/133116_creeks31x.html> (accessed December 12, 2007).

Mulady, Kathy. 2003c. Nickels Eases His Stance on Northgate. *Seattle Post-Intelligencer*, November 18, <http://www.seattlepi.com/local/148778_northgate18.html> (accessed December 12, 2007).

Mulady, Kathy. 2003d. Northgate Compromise Plan Wins Approval. *Seattle Post-Intelligencer*, December 9, <http://www.seattlepi.com/local/151667_northgate09.html> (accessed December 12, 2007).

Mulady, Kathy. 2004. Thornton Creek May See Daylight Again. *Seattle Post-Intelligencer*, June 8, <http://www.seattlepi.com/local/176819_creek08.html> (accessed December 12, 2007).

Mumford, Lewis. 1956. The Natural History of Urbanization. In *Man's Role in Changing the Face of the Earth*, ed. William L. Thomas Jr., 382–398. Chicago: University of Chicago Press.

Murdoch, Jonathan. 1997. Towards a Geography of Heterogeneous Associations. *Progress in Human Geography* 21 (3): 321–337.

Murdoch, Jonathan. 1998. The Spaces of Actor-Network Theory. *Geoforum* 29 (4): 357–374.

Murdoch, Jonathan. 2001. Ecologising Sociology: Actor-Network Theory, Co-construction and the Problem of Human Exemptionalism. *Sociology* 35 (1): 111–133.

Murdoch, Jonathan. 2006. *Post-Structuralist Geography: A Guide to Relational Space*. Thousand Oaks, Calif.: Sage Publications.

Nabbefeld, Joe. 1999. Up a Creek: Northgate Project Faces ESA Issues. *Puget Sound Business Journal*, April 11, <http://www.bizjournals.com/seattle/stories/1999/04/12/story2.html>

Ndubisi, Forster. 1997. Landscape Ecological Planning. In *Ecological Design and Planning*, ed. George F. Thompson and Frederick R. Steiner, 9–44. New York: John Wiley and Sons.

Novotny, Vladimir. 2003. *Water Quality: Diffuse Pollution and Watershed Management*. New York: John Wiley and Sons.

Nye, David E., ed. 1999. *Technologies of Landscape: From Reaping to Recycling*. Amherst: University of Massachusetts Press.

Nye, David E. 2003a. *America as Second Creation: Technology and Narratives of New Beginnings*. Cambridge, Mass.: MIT Press.

Nye, David E. 2003b. Technology, Nature, and American Origin Stories. *Environmental History* 8 (1): 8–24.

O'Donnell, Arthur J. 2004. *In the City of Neighborhoods: Seattle's History of Community Activism and Non-Profit Survival Guide*. New York: iUniverse.

Oden, Michael. 1997. From Assembly to Innovation: The Evolution and Current Structure of Austin's High Tech Economy. *Planning Forum* 3:14–30.

Oliver, Stuart. 2000. The Thames Embankment and the Disciplining of Nature in Modernity. *Geographical Journal* 166 (3): 227–238.

Olmsted, Frederick Law, Jr. 1916. Introduction. In *City Planning: A Series of Papers Presenting the Essential Elements of the City Plan*, ed. J. Nolen, 1–19. New York: D. Appleton and Co.

Open Space Seattle 2100. 2008. Open Space Seattle 2100, <http://www.open2100 .org> (accessed May 8, 2008).

Orum, Anthony M. 1987. *Power, Money and the People: The Making of Modern Austin.* Austin: Texas Monthly Press.

Pasternak, Scott. 1995. "Funding and Maintaining Urban Stormwater Programs: Effective and Efficient Use of Stormwater Utilities." Master's report, Community and Regional Planning, University of Texas at Austin.

Pellow, David Naguib. 2002. *Garbage Wars: The Struggle for Environmental Justice in Chicago.* Cambridge, Mass.: MIT Press.

Pickett, S. T. A., M. L. Cadenasso, J. M. Grove, C. H. Nilon, R. V. Pouyat, W. C. Zipperer, and R. Costanza. 2001. Urban Ecological Systems: Linking Terrestrial Ecological, Physical, and Socioeconomic Components of Metropolitan Areas. *Annual Review of Ecology and Systematics* 32:127–157.

Pieranunzi, Danielle Deborah. 2006. "Conservation Developments: Transitions Toward Sustainable Landscapes and Societies." Master's thesis, Sustainable Design, University of Texas at Austin.

Pincetl, Stephanie. 2004. The Preservation of Nature at the Urban Fringe. In *Up Against the Sprawl: Public Policy and the Making of Southern California,* ed. Jennifer Wolch, Manuel Pastor, and Peter Dreier, 225–251. Minneapolis: University of Minnesota Press.

Piper's Creek Watershed Management Committee. 1990. *Piper's Creek Watershed Action Plan for the Control of Nonpoint Source Pollution.* Seattle: n.p.

Pipkin, Turk. 1995. *Born of the River: The Colorado River and the LCRA.* Austin: Softshoe Publishing.

Pipkin, Turk, and Marshall Frech, eds. 1993. *Barton Springs Eternal: The Soul of a City.* Austin: Softshoe Publishing.

Price, Asher. 2006a. City Delivers Green Win in East Austin. *Austin American Statesman,* November 30, <http://www.statesman.com/news/content/news/ stories/local/11/30/30oaksprings.html> (accessed November 30, 2006).

Price, Asher. 2006b. Hill County Development Bringing Big Environmental Concerns. *Austin American-Statesman,* December 24, <http://www.statesman.com/ news/content/news/stories/local/12/24/24hcpart2.html> (accessed December 24, 2006).

Price, Asher. 2007. SOS Examining Options. *Austin American-Statesman,* April 29, <http://www.statesman.com/news/content/news/stories/local/04/29/ 29sos.html> (accessed April 29, 2007).

Prugh, Thomas, Robert Costanza, and Herman Daly. 2000. *The Local Politics of Global Sustainability.* Washington, D.C.: Island Press.

Public Employees for Environmental Responsibility. 2008. "Local Control"— Texas Style. <http://www.txpeer.org/Bush/Local_Control.html> (accessed May 15, 2008).

Pursell, Carroll. 1993. The Rise and Fall of the Appropriate Technology Movement in the United States, 1965–1985. *Technology and Culture* 34 (3): 629–637.

Raban, Jonathan. 1993. The Next Last Frontier: A Newcomer's Journey Through the Pacific Northwest. *Harper's Magazine* 272 (2): 30–48.

Raban, Jonathan. 2001. Battleground of the Eye. *Atlantic Monthly* 287 (3): 40–52.

Reeves, Kimberly. 2007. The Light at the Beginning of the Waller Tunnel. *Austin Chronicle*, November 2, <http://www.austinchronicle.com/gyrobase/Issue/story?oid=oid:556556> (accessed November 2, 2007).

Reid, Herbert, and Betsy Taylor. 2003. John Dewey's Aesthetic Ecology of Public Intelligence and the Grounding of Civic Environmentalism. *Ethics and the Environment* 8 (1): 74–92.

Reisner, Marc. 1987. *Cadillac Desert: The American West and its Disappearing Water*. New York: Penguin Books.

Reynolds, Terry S., ed. 1991. *The Engineer in America*. Chicago: University of Chicago Press.

Richards, Kent D. 1981. In Search of the Pacific Northwest: The Historiography of Oregon and Washington. *Pacific Historical Review* 50 (4): 415–443.

Robbins, William G. 2001. Nature's Northwest. In *Search of a Pacific Region, in The Great Northwest: The Search for Regional Identity*, ed. William G. Robbins, 158–182. Corvallis, Ore.: Oregon State University Press.

Roseland, Mark. 2006. *Toward Sustainable Communities: Resources for Citizens and Their Governments*. Rev. ed. Gabriola Island, B.C., Canada: New Society Publishers.

Ross, Lauren, and Jeanine Sih. 1997. *Protecting the Edwards Aquifer: A Scientific Consensus*. Austin: n.p.

Rottle, Nancy, and Brice Maryman. 2006. Strategies for a Green Future. *Landscape Architecture* 96 (11): 58–65.

Rubin, Charles T. 2002. Civic Environmentalism. In *Democracy and the Claims of Nature: Critical Perspectives for a New Century*, ed. Ben A. Minteer and Bob Pepperman Taylor, 335–351. Lanham, Md.: Rowman and Littlefield.

Rybczynski, Witold. 1999. *A Clearing in the Distance: Frederick Law Olmsted and America in the 19th Century*. New York: Scribner.

Ryn, Sim Van der, and Stuart Cowan. 1996. *Ecological Design*. Washington, D.C.: Island Press.

Sale, Kirkpatrick. 1985. *Dwellers in the Land: The Bioregional Vision*. San Francisco: Sierra Club Books.

Sale, Roger. 1976. *Seattle: Past to Present*. Seattle: University of Washington Press.

Sanders, Jeffrey Craig. 2005. "Inventing Ecotopia: Nature, Culture, and Urbanism in Seattle, 1960–2000." Doctoral dissertation, History, University of New Mexico, Albuquerque.

Sansalone, John J., and Jonathan Hird. 2003. Treatment of Stormwater Runoff from Urban Pavement and Roadways. In *Wet-Weather Flow in the Urban Water-*

shed: Technology and Management, ed. Richard Field and Daniel Sullivan, 141–185. New York: Lewis Publishers.

Sato, Mike. 1997. *The Price of Taming a River: The Decline of Puget Sound's Duwamish/Green Waterway*. Seattle: The Mountaineers.

Schauman, Sally. 1997. Beyond Stewardship Toward Partnership. In *Ecological Design and Planning*, ed. George F. Thompson and Frederick R. Steiner, 239–262. New York: John Wiley and Sons.

Schauman, Sally, and Sandra Salisbury. 1998. Restoring Nature in the City: Puget Sound Experiences. *Landscape and Urban Planning* 42:287–295.

Schell, Paul. 1999. Saving Salmon May Mean Saving Ourselves. *Seattle Post-Intelligencer*, March 14, <http://www.seattlepi.com/opinion/focusschell.shtml> (accessed December 12, 2007).

Schell, Paul, and John Hamer. 1995. Cascadia: The New Binationalism of Western Canada and the U.S. Pacific Northwest. In *Identities of North America: The Search for Community*, ed. Robert L. Earle and John D. Wirth, 140–156. Stanford: Stanford University Press.

Schlossberg, David. 2007. *Defining Environmental Justice*. New York: Oxford University Press.

Schueler, Thomas, and Richard Claytor. 1997. Impervious Cover as a Urban Stream Indicator and a Watershed Management Tool. In *Effects of Watershed Development and Management on Aquatic Ecosystems*, ed. Larry A. Roesner, 513–529. New York: ASCE.

Schultz, Stanley K. 1989. *Constructing Urban Culture: American Cities and City Planning, 1800–1920*. Philadelphia: Temple University Press.

Schultz, Stanley K., and Clay McShane. 1978. To Engineer the Metropolis: Sewers, Sanitation, and City Planning in Late-Nineteenth-Century America. *Journal of American History* 65 (2): 389–411.

Schulze, Peter C., ed. 1996. *Engineering Within Ecological Constraints*. Washington, D.C.: National Academy Press.

Schwantes, Carlos A. 1989. *The Pacific Northwest: An Interpretive History*. Lincoln: University of Nebraska Press.

Sclove, Richard. 1995. *Democracy and Technology*. New York: Guilford Press.

Sclove, Richard. 2000. Town Meetings on Technology: Consensus Conferences as Democratic Participation. In *Science, Technology, and Democracy*, ed. Daniel L. Kleinman, 33–48. Albany: State University of New York Press.

Seattle Convention and Visitors Bureau. 2006. Metronatural™ Press Kit, <http://www.visitseattle.org/Meetings-And-Conventions/Planning-Toolkit/Destination-Branding.aspx> (accessed February 8, 2008).

Seattle Public Utilities. 2005. *City of Seattle 2004 Comprehensive Drainage Plan*. Seattle: Seattle Public Utilities.

Seattle Public Utilities. 2007. *State of the Waters 2007*. Seattle: Seattle Public Utilities.

Seattle Public Utilities. 2008. Natural Drainage Systems website, <http://www
.seattle.gov/util/About_SPU/Drainage_&_Sewer_System/GreenStormwater
Infrastructure/NaturalDrainageProjects/index.htm> (accessed March 12, 2008).

Seattle Times. 1999. Urban Design / Northgate Mall—A New Northgate Can
Revitalize Community. *Seattle Times*, April 19, <http://community.seattletimes
.nwsource.com/archive/?date=19990419&slug=2955918> (accessed December 12, 2007).

Seattle Times. 2002. Initiative Filed to Open Creeks. *Seattle Times*, June 14,
<http://community.seattletimes.nwsource.com/archive/?date=20020614&slug
=dige14m> (accessed December 12, 2007).

Selden, Jonathan. 2006. Council May Wade into Aquifer Debate. *Austin Business
Journal*, October 15, <http://www.bizjournals.com/austin/stories/2006/10/16/
story1.html> (accessed October 16, 2006).

Seney, Donald Bradley. 1975. "The Development of Water, Sewer and Fire Service
in Seattle and King County, Washington: 1851–1966." Doctoral dissertation,
Political Science/Public Affairs, University of Washington, Seattle.

Sevcik, Edward A. 1992. Selling the Austin Dam: A Disastrous Experiment in
Encouraging Growth. *Southwestern Historical Quarterly* 96 (2): 215–240.

Sheldon, Kara C. 2000. "Combating Suburban Sprawl in Austin: The Foundation
for an Effective Smart Growth Program." Master's report, Public Affairs,
University of Texas at Austin.

Shevory, Kristina. 2007. Battling to Keep the Country in the Texas Hill Country.
New York Times, July 8, <http://www.nytimes.com/2007/07/08/realestate/08nati
.html> (accessed November 24, 2007).

Shrake, Bud. 1974. The Screwing Up of Austin. *Texas Observer* 66:26–27.

Shuster, W. D., J. Bonta, H. Thurston, E. Warnemuende, and D. R. Smith. 2005.
Impacts of Impervious Surface on Watershed Hydrology: A Review. *Urban Water
Journal* 2 (4): 263–275.

Shutkin, William A. 2000. *The Land that Could Be: Environmentalism and Democracy in the Twenty-First Century*. Cambridge, Mass.: MIT Press.

Shutkin, William. 2005. From Pollution Control to Place Making: The Role of
Environmental Regulation in Creating Communities of Place. In *Regulating Place:
Standards and the Shaping of Urban America*, ed. Eran Ben-Joseph and Terry S.
Szold, 253–269. New York: Routledge.

Simpson, Buster. 2004. Perpendicular and Parallel Streetscape Stories. In *Belltown
Paradise/Making Their Own Plans*, ed. Brett Bloom and Ava Bromberg, 40–60.
Chicago: WhiteWalls.

Sirianni, Carmen, and Lewis Friedland. 2001. *Civic Innovation in America: Community Empowerment, Public Policy, and the Movement for Civic Renewal*. Berkeley: University of California Press.

Smith, Elizabeth. 2001. *Austin's Evolution: University Town to High Tech Center*.
Austin: Greater Austin Chamber of Commerce.

Smith, Mark J., and Piya Pangsapa. 2008. *Environment and Citizenship: Integrating Justice, Responsibility and Civic Engagement.* London: Zed Books.

Southworth, Michael, and Eran Ben-Joseph. 2003. *Streets and the Shaping of Towns and Cities.* Washington, D.C.: Island Press.

Spirn, Anne Whiston. 1984. *The Granite Garden: Urban Nature and Human Design.* New York: Basic Books.

Spirn, Anne Whiston. 1996. Constructing Nature: The Legacy of Frederick Law Olmsted. In *Uncommon Ground: Rethinking the Human Place in Nature,* ed. William Cronon, 91–113. New York: W. W. Norton.

Spirn, Anne Whiston. 1998. *The Language of Landscape.* New Haven, Conn.: Yale University Press.

Spirn, Anne Whiston. 2005. Restoring Mill Creek: Landscape Literacy, Environmental Justice and City Planning and Design. *Landscape Research* 30 (3): 395–413.

Steinbrueck, Victor. 1962. *Seattle Cityscape.* Seattle: University of Washington Press.

Steinbrueck, Victor. 1973. *Seattle Cityscape #2.* Seattle: University of Washington Press.

Steiner, Frederick R. 2002. *Human Ecology: Following Nature's Lead.* Washington, D.C.: Island Press.

Steiner, Frederick R. 2005. The Woodlands: Retrospect and Prospect. *Cite* 65: 35–41.

Steiner, Frederick R. 2006. Metropolitan Resilience: The Role of Universities in Facilitating a Sustainable Metropolitan Future. In *Toward a Resilient Metropolis: The Role of State and Land Grant Universities in the 21st Century,* ed. Arthur C. Nelson, Barbara L. Allen, and David L. Trauger, 1–18. Alexandria, Va.: Metropolitan Institute at Virginia Tech.

Stilgoe, Jack, Alan Irwin, and Kevin Jones. 2006. *The Received Wisdom: Opening Up Expert Advice.* London: Demos.

Summerton, Jane, ed. 1994. *Changing Large Technical Systems.* Boulder: Westview Press.

SustainLane. 2008. 2008 SustainLane US City Rankings, <http://www.sustainlane.com/us-city-rankings/> (accessed January 21, 2011).

SvR Design Company. 2009. *Thornton Creek Water Quality Channel, Final Report.* Seattle: Seattle Public Utilities.

Swearingen, Scott. 1997. Sacred Space in the City: Community Totems and Political Contests. *Research in Community Sociology* 7:141–169.

Swearingen, William Scott, Jr. 2010. *Environmental City: People, Place, Politics, and the Meaning of Modern Austin.* Austin: University of Texas Press.

Swyngedouw, Erik. 2004. *Social Power and the Urbanization of Water: Flows of Power.* New York: Oxford University Press.

Swyngedouw, Erik. 2006. Circulations and Metabolisms: (Hybrid) Natures and (Cyborg) Cities. *Science as Culture* 15 (2): 105–121.

Swyngedouw, Erik. 2007. Impossible "Sustainability" and the Postpolitical Condition. In *The Sustainable Development Paradox: Urban Political Economy in the United States and Europe*, ed. Rob Krueger and David Gibbs, 13–40. New York: Guilford Press.

Swyngedouw, Erik, and Maria Kaika. 2000. The Environment of the City . . . or the Urbanization of Nature. In *A Companion to the City*, ed. Gary Bridge and Sophie Watson, 567–580. Oxford: Blackwell.

Tarr, Joel A. 1979. The Separate vs. Combined Sewer Problem: A Case Study in Urban Technology Design Choice. *Journal of Urban History* 5 (3): 308–339.

Tarr, Joel A. 1988. Sewerage and the Development of the Networked City in the United States, 1850–1930. In *Technology and the Rise of the Networked City in Europe and America*, ed. Joel A. Tarr and Gabriel Dupuy, 159–185. Philadelphia: Temple University Press.

Tarr, Joel A., and Josef W. Konvitz. 1987. Patterns in the Development of the Urban Infrastructure. In *American Urbanism: A Historiographical Review*, ed. Howard Gillette Jr. and Zane L. Miller, 195–226. New York: Greenwood Press.

Tarr, Joel A., James McCurley, III, Francis C. McMichael, and Terry Yosie. 1984. Water and Wastes: A Retrospective Assessment of Wastewater Technology in the United States, 1800–1932. *Technology and Culture* 25 (2): 226–263.

Taylor, Joseph E., III. 1999. *Making Salmon: An Environmental History of the Northwest Fisheries Crisis*. Seattle: University of Washington Press.

Thayer, Robert L., Jr. 1994. *Gray World, Green Heart: Technology, Nature, and the Sustainable Landscape*. New York: John Wiley and Sons.

Thompson, George F., and Frederick R. Steiner, eds. 1997. *Ecological Design and Planning*. New York: John Wiley and Sons.

Thomson, R. H. 1950. *That Man Thomson*. Seattle: University of Washington Press.

Thornton Creek Watershed Management Committee. 2000. *Thornton Creek Watershed Characterization Report*. Seattle: Seattle Public Utilities.

Thornton Place. 2011. Thornton Place website, <http://www.thornton-place.com> (accessed January 21, 2011).

Thrift, Nigel. 1999. Steps to an Ecology of Place. In *Human Geography Today*, ed. Doreen Massey, John Allen, and Philip Sarre, 295–322. Malden, Mass.: Polity Press.

Thrift, Nigel. 2007. *Non-representational Theory: Space, Politics, Affect*. New York: Routledge.

Thrush, Coll. 2006. City of the Changers: Indigenous People and the Transformation of Seattle's Watersheds. *Pacific Historical Review* 75 (1): 89–117.

Thrush, Coll. 2007. *Native Seattle: Histories from the Crossing-Over Place*. Seattle: University of Washington Press.

True, Kathryn. 2005. The Poetry of Longfellow Creek. *Seattle Times*, August 18, <http://seattletimes.nwsource.com/html/outdoors/2002443303_nwwcreek18 .html> (accessed December 12, 2007).

Tubbs, Donald W. 1974. *Landslides in Seattle. State of Washington, Department of Natural Resources*. Olympia: Department of Natural Resources.

U.S. Census. 2000. U.S. Census, <http://www.census.gov> (accessed May 15, 2008).

U.S. Congress. 1972. Clean Water Act. 33 U.S.C. §1251 et seq.

U.S. Department of Interior. 1997. Endangered and Threatened Wildlife and Plants; Final Rule to List the Barton Springs Salamander as Endangered. *Federal Register* 62 (83) (April 30): 23377–23392.

U.S. Environmental Protection Agency. 1983. *Final Report*. Vol. 1. Results of the Nationwide Urban Runoff Program. Washington, D.C.: U.S. Environmental Protection Agency.

U.S. Environmental Protection Agency. 1997. *Community-Based Environmental Protection: A Resource Book for Protecting Ecosystems and Communities*. Washington, D.C.: U.S. Environmental Protection Agency.

U.S. Environmental Protection Agency. 2000. *Low Impact Development (LID): A Literature Review*. U.S. EPA report EPA-841-B-00–005. Washington, D.C.: U.S. Environmental Protection Agency.

U.S. Environmental Protection Agency. 2008. EPA's Clean Water Act Recognition Awards, <http://water.epa.gov/scitech/wastetech/cwa-awards-history.cfm> (accessed April 21, 2008).

Vileisis, Ann. 2000. *Discovering the Unknown Landscape: A History of America's Wetlands*. Washington, D.C.: Island Press.

Vogel, Mary. 2006. Moving Toward High-Performance Infrastructure. *Urban Land* 65 (10): 73–79.

Waldheim, Charles, ed. 2006. *The Landscape Urbanism Reader*. New York: Princeton Architectural Press.

Walesh, Stuart G. 1989. *Urban Surface Water Management*. New York: John Wiley and Sons.

Walz, Karen. 2005. A Sleeper: Regional Planning in Texas. *Planning* 71:16–17.

Wassenich, Rob. 2007. *Keep Austin Weird: A Guide to the Odd Side of Town*. Atglen, Pa.: Schiffer Publishing.

Weber, Erwin L. 1924. *In the Zone of Filtered Sunshine: Why the Pacific Northwest is Destined to Dominate the Commercial World*. Seattle: Pacific Northwest Publishing Company.

Weibel, S. R., R. J. Anderson, and R. L. Woodward. 1964. Urban Land Runoff as a Factor in Stream Pollution. *Water Pollution Control Federation Journal* 36 (7): 914–924.

Weiss, Peter T., John S. Gulliver, and Andrew J. Erickson. 2007. Cost and Pollutant Removal of Storm-Water Treatment Practices. *Journal of Water Resources Planning and Management* 133 (3): 218–229.

Whatmore, Sarah. 1999. Hybrid Geographies: Rethinking the "Human" in Human Geography. In *Human Geography Today*, ed. Doreen Massey, John Allen, and Philip Sarre, 22–39. Malden, Mass.: Polity Press.

Whatmore, Sarah. 2002. *Hybrid Geographies: Natures, Cultures, Spaces*. Thousand Oaks, Calif.: Sage Publications.

White, Richard. 1995. *The Organic Machine*. New York: Hill and Wang.

White, Richard. 1996. The Nature of Progress: Progress and the Environment. In *Progress: Fact or Illusion?*, ed. Leo Marx and Bruce Mazlish, 121–140. Ann Arbor: University of Michigan Press.

Wiland, Harry, and Dale Bell. 2006. *Edens Lost and Found: How Ordinary Citizens are Restoring Our Great American Cities*. White River Junction, Vt.: Chelsea Green Publishing Company.

Williams, David B. 2005. *The Street-Smart Naturalist: Field Notes from Seattle*. Portland, Ore.: Westwinds Press.

Williams, Raymond. 1972. Ideas of Nature. In *Ecology, the Shaping Enquiry*, ed. Jeremy Benthall, 146–164. London: Longman.

Willis, Lance. 2006. The Stormwater Story. *Stormwater* 7 (7): 12–22.

Wilson, Mark Griswold, and Emily Roth. 2006. Urban Natural Areas. In *Restoring the Pacific Northwest: The Art and Science of Ecological Restoration in Cascadia*, ed. Dean Apostol and Marcia Sinclair, 279–297. Washington, D.C.: Island Press.

Winterbottom, Daniel. 2005. Applying the Theory: Design/Build Models for Water Harvesting and Watershed Awareness. In *Facilitating Watershed Management: Fostering Awareness and Stewardship*, ed. Robert L. France, 231–251. Lanham, Md.: Rowman and Littlefield.

Wolch, Jennifer. 2007. Green Urban Worlds. *Annals of the Association of American Geographers* 97 (2): 373–384.

Worster, Donald. 1990. The Ecology of Order and Chaos. *Environmental History Review* 14 (1/2): 1–18.

Worster, Donald. 1993. *The Wealth of Nature: Environmental History and the Ecological Imagination*. New York: Oxford University Press.

Wurth, Albert H., Jr. 1996. Why Aren't All Engineers Ecologists? In *Engineering within Ecological Constraints*, ed. Peter C. Schulze, 129–140. Washington, D.C.: National Academy Press.

Yocom, Kenneth. 2007. "Building Watershed Narratives: Two Case Studies of Urban Streams in Seattle, Washington." Doctoral dissertation, College of Built Environment, University of Washington, Seattle.

Young, Bob. 2003. Compromise to Start Upgrade of Northgate. *Seattle Times*, December 9, <http://community.seattletimes.nwsource.com/archive/?date=2003 1209&slug=northgate09m> (accessed December 12, 2007).

Young, Bob, and Jake Batsell. 2003. Nickels Wants to Repeal Northgate Mall Plan. *Seattle Times*, March 18, <http://community.seattletimes.nwsource.com/archive/?date=20030318&slug=northgate18m0> (accessed December 9, 2003).

Austin Interviewees

Aus60 Local geologist
Aus61 Municipal staff member
Aus62 Environmental activist
Aus63 Municipal staff member
Aus64 Lower Colorado River Authority engineer
Aus65 Consulting engineer
Aus66 Environmental activist
Aus67 Environmental activist
Aus69 Environmental activist
Aus71 Environmental activist
Aus72 Municipal staff member
Aus73 Environmental activist
Aus74 City Hall insider
Aus75 Municipal staff member

Seattle Interviewees

Sea61 Municipal staff member
Sea62 Municipal staff member
Sea63 Landscape architect
Sea701 Consulting engineer
Sea702 Municipal staff member
Sea707 Municipal staff member
Sea708 City Hall insider
Sea709 Environmental activist
Sea711 Neighborhood activist
Sea714 Municipal staff member
Sea81 Municipal staff member
Sea82 University of Washington faculty member
Sea84 Municipal staff member
Sea85 Municipal staff member

Index

Urban and Industrial Environments

Series editor: Robert Gottlieb, Henry R. Luce Professor of Urban and Environmental Policy, Occidental College

Maureen Smith, *The U.S. Paper Industry and Sustainable Production: An Argument for Restructuring*

Keith Pezzoli, *Human Settlements and Planning for Ecological Sustainability: The Case of Mexico City*

Sarah Hammond Creighton, *Greening the Ivory Tower: Improving the Environmental Track Record of Universities, Colleges, and Other Institutions*

Jan Mazurek, *Making Microchips: Policy, Globalization, and Economic Restructuring in the Semiconductor Industry*

William A. Shutkin, *The Land That Could Be: Environmentalism and Democracy in the Twenty-First Century*

Richard Hofrichter, ed., *Reclaiming the Environmental Debate: The Politics of Health in a Toxic Culture*

Robert Gottlieb, *Environmentalism Unbound: Exploring New Pathways for Change*

Kenneth Geiser, *Materials Matter: Toward a Sustainable Materials Policy*

Thomas D. Beamish, *Silent Spill: The Organization of an Industrial Crisis*

Matthew Gandy, *Concrete and Clay: Reworking Nature in New York City*

David Naguib Pellow, *Garbage Wars: The Struggle for Environmental Justice in Chicago*

Julian Agyeman, Robert D. Bullard, and Bob Evans, eds., *Just Sustainabilities: Development in an Unequal World*

Barbara L. Allen, *Uneasy Alchemy: Citizens and Experts in Louisiana's Chemical Corridor Disputes*

Dara O'Rourke, *Community-Driven Regulation: Balancing Development and the Environment in Vietnam*

Brian K. Obach, *Labor and the Environmental Movement: The Quest for Common Ground*

Peggy F. Barlett and Geoffrey W. Chase, eds., *Sustainability on Campus: Stories and Strategies for Change*

Steve Lerner, *Diamond: A Struggle for Environmental Justice in Louisiana's Chemical Corridor*

Jason Corburn, *Street Science: Community Knowledge and Environmental Health Justice*

Peggy F. Barlett, ed., *Urban Place: Reconnecting with the Natural World*

David Naguib Pellow and Robert J. Brulle, eds., *Power, Justice, and the Environment: A Critical Appraisal of the Environmental Justice Movement*